D1664969

Handbook of Machine Tools
Volume 4

HANDBOOK OF MACHINE TOOLS

Handbook of Machine Tools
Volume 4
Metrological Analysis and Performance Tests

Manfred Weck
Lehrstuhl für Werkzeugmaschinen
Laboratorium für Werkzeugmaschinen und Betriebslehre
Aachen, Germany

Translated from the original German by
H. BIBRING
Senior Lecturer, Middlesex Polytechnic
London, U.K.

A Wiley Heyden Publication

JOHN WILEY & SONS
Chichester · New York · Brisbane · Toronto · Singapore

Title of German edition:
Werkzeugmaschinen, Band 1:
Maschinenarten, Bauformen und Anwendungsbereiche
von Prof. Dr.-Ing. Manfred Weck
Laboratorium für Werkzeugmaschinen und Betriebslehre
der Rheinisch-Westfälischen Technischen Hochschule, Aachen
Zweite, neubearbeitete Auflage

© VDI-Verlag GmbH, Düsseldorf, 1980

Library of Congress Cataloging in Publication Data:

Weck, Manfred, 1937–
   Handbook of machine tools.

    Translation of: Werkzeugmaschinen.
    'A Wiley–Heyden publication.'
    Includes bibliographies and indexes.
    Contents: v. 1. Types of machines, forms of construction,
and applications—v. 2. Construction and mathematical
analysis—v. 3. Automation and controls—
[etc.]
    1. Machine-tools. I. Title.
TJ1185.W382313   1984   621.9'02   83-23483
ISBN 0 471 26226 9 (set)
ISBN 0 471 26225 0 (Vol. 4)

British Library Cataloguing in Publication Data

Weck, Manfred
   Handbook of machine tools.—(A Wiley–Heyden publication)
   Vol. 4: Metrological analysis and performance tests
   1. Machine-tools
   I. Title   II. Bibring, H.
   III. Werkzeugmaschien Bd 4. *English*
   621.9'02   TJ1185

   ISBN 0 471 26226 9 (set)
   ISBN 0 471 26225 0 (Vol. 4)

Set in VIP Times by Preface Ltd, Salisbury, Wilts.
and printed in Great Britain by Page Brothers (Norwich) Ltd

CONTENTS

FOREWORD

Machine tools are among the most important means of production for the metal-working industries. Without the development of this type of machine, the high living standards of the present time would be unthinkable. In some of the most highly industrialized nations, approximately 10% of all machines built are machine tools, and about 10% of the work force in machine manufacture is concerned with machine tools.

The form of construction and degree of automation of machine tools are just as varied as their fields of application. The wide field that is embraced ranges from casting and forming, through cutting (including abrasion), to welding and assembly, and is limited only by the degree of technological development. According to the component to be produced and the quantities involved, these machines are variedly automated with a greater or lesser degree of flexibility. Thus, single-purpose or special-purpose machines are available to the user, as are universal machines offering a wide range of applications.

As a result of the increased demands in both performance and precision, the manufacturer of these machines has to determine the optimum design of the individual machine components. To this end, he requires an in-depth knowledge of the capabilities and capacities of the components and elements of his machine.

Today, as a result of the availability of a very comprehensive library of programs, the necessary calculations can be carried out with the aid of computers. Metrological analysis and objective performance tests have opened up the possibilities of critically determining the machine's capabilities and degree of accuracy of production, as well as determining its geometric, kinematic, static, dynamic, thermal and acoustic properties, and from this information, to initiate any improvements thought to be necessary.

The constant tendency towards further automation of machine tools has resulted in the development of a wide range of alternative controls. In recent years, the development of electronics has had a marked effect on machine-tool controls. Microprocessors and process calculators have made possible control techniques which were previously unthinkable. Mechanization and automation have also been applied in the area of material transportation and the feeding of the machines with material. The work done in these fields has resulted in the availability of transfer machines for the mass production industry and for medium- and small-quantity manufacture, and has led to the availability of flexible production systems.

The four volumes published under the general title '*Handbook of Machine Tools*' are intended for the use of students in the field of production engineering, as well as established specialists in the field who require to keep abreast of the constantly developing field of this branch of machine construction. A further aim of these volumes is to assist users in the choice of suitable machines and their controls. For the machine-tool manufacturer, optimum layouts of machine components, drives and controls are indicated, and it is shown how improvements can be achieved as a result of metrological analysis and objective performance tests. Their content is derived from and well validated by lectures given at the Technical University for the Rhineland and Westphalia in Aachen.

MANFRED WECK

PREFACE

The quality control of machine tools and the acceptance of the machine by the user when delivered require that all important features are defined and evaluated. These demand objective techniques for determining and describing the appropriate parameters.

The problem which arises is that practically all machine characteristics are influenced by a large number of periodic or systematic and random variables. In order to arrive at a scientific judgement, it is necessary therefore to recognize the individual variants and to ensure that they are constant when comparing appraisals of machine tools.

In addition to the definitions of the characteristics, performance accuracies, capabilities and effects on the environment of machines, fundamental relationships are presented in this volume, the important parameters which influence the machine performance are discussed and the current knowledge of the metrological technology is indicated.

To these ends, there are a number of references to the extent to which tests of any particular machine characteristic are laid down in standards or guidelines. Moreover, in relation to the metrological technology necessary for determination of the machine characteristics, procedures are described which ensure recognition of the weaknesses in machines with respect to the individual characteristics in order for steps to be taken which will lead to their improvement.

This volume has been produced with the collaboration of my colleagues, Dipl.-Ing. W. Möllers, Dipl.-Ing. K. Klumpers, Dipl.-Ing. E.-K. Prössler, Dipl.-Ing. W. Brey, as well as Dr.-Ing. K. Teipel, who also took responsibility for the co-ordination of the individual sections of this book. I would like to take this opportunity to express my sincere gratitude for their readiness to assist in this work. I am also indebted to Dr.-Ing. W. Borchert from the VDI publishing house for the painstaking examination of the manuscripts. Similarly, my appreciation is recorded to A. Hümmler for his careful proof-reading of the English manuscript.

MANFRED WECK

NOMENCLATURE AND ABBREVIATIONS

Upper case letters

$[A]_{<}$	—	Matrix of angular errors
B	μm	Maximum deflection
D	—	Damping ratio
$E_x(j\omega)$	—	Complex energy spectrum of input to system
$E_y(j\omega)$	—	Complex energy spectrum of output from system
F	mm	Length of elastic deflection
F	N	Force
$F(j\omega)$	N	Force spectrum
$\hat{F}$	N	Peak value of a sinusoidal force
$G_d(j\omega)$	μm N^{-1}	Directional flexibility–frequency characteristic
$G_{ij}(j\omega)$	μm N^{-1}	Flexibility–frequency characteristic
		$i \stackrel{\wedge}{=}$ direction of force vector
		$j \stackrel{\wedge}{=}$ direction of distance vector
$G_0(j\omega)$	—	Frequency characteristic of open circuit
$[G]$	μm N^{-1}	Matrix of flexibility–frequency characteristic of machine
H	—	Cumulative repetition
I	W m^{-2}	Sound intensity
Im	—	Imaginary portion
I_0	W m^{-2}	Reference sound intensity
K_i	dB(A)	Sound level increase
K_1	dB(A)	Correction value for external noise
K_2	dB(A)	Correction value for ambient effects
L, L_p	dB	Sound pressure level
L_A	dB(A)	Corrected sound pressure level
L_m	dB(A)	Mean sound pressure level
L_{mT}	dB(A)	Mean sound pressure level according to the cyclic maximum level technique
L_r	dB(A)	Evaluation sound pressure level
L_s	dB(A)	Measuring surface value of sound pressure level
L_W	dB	Acoustic power level
L_{WA}	dB(A)	Corrected acoustic power level
P	W	Power, acoustic power
P_0	W	Reference acoustic power
P_0	—	Reference point in a machine coordinate system

R	—	Correlation coefficient
Re	—	Real portion
S	m^2	Measuring area
S_0	m^2	Reference measuring area
$S_x(j\omega), S_x^*(j\omega)$	—	Complex power spectrum of input to system; complex conjugate
$S_y(j\omega), S_y^*(j\omega)$	—	Complex power spectrum of output from system; complex conjugate
$S_{xx}(\omega), S_{yy}(\omega)$	—	Auto-power spectrum (real) from input and output of system
$S_{xy}(j\omega)$	—	Cross-power spectrum (complex) from input and output of system
T	N m	Torque
T	s	Time, period
T_d	s	Dead-time
T_i	s	Part of time
T_r	s	Evaluation period
U	μm	Uncertainty
V	m^3	Volume
$\{X\}$	—	Positional vector in the machine coordinate system

Lower case letters

a	mm	Depth of cut
a	μm	Deflection
b	mm	Width of cut
c	N s mm^{-1}	Damping coefficient
c	m s^{-1}	Sound velocity
d	mm	Diameter
d_{ij}	—	Directional factor
f	Hz	Frequency
f_m	Hz	Mean frequency
f_n	Hz	Natural frequency
f_l	Hz	Lower limiting frequency
f_u	Hz	Upper limiting frequency
f_R	Hz	Resonance frequency
h	—	Relative repetition
i	μm	Tolerance unit
k	N mm^{-1}	Stiffness coefficient (spring coefficiency)
k_{cb}	N μm^{-2}	Specific dynamic cutting coefficient
l	mm	Length
m	kg	Mass
n	rev min^{-1}	Rotational speed
p	N m^{-2}	Sound pressure

p_0	N m^{-2}	Reference sound pressure
s	mm	Feed
s_z	mm	Feed per cutting edge
s	—	Standard deviation
s^2	—	Variance
t	s	Time
u	mm	Dynamic cutting thickness
v	m s^{-1}	Cutting velocity
v	m s^{-1}	Acoustic velocity
x	μm	Displacement, distance
$\hat{x}$	μm	Peak value of a sinusoidal displacement
x_z	μm	Amplite of the zth period of a decaying sinusoidal vibration
x, y, z	—	Co-ordinate directions
z	—	Number of cutting edges, number of teeth

Greek letters

β	deg	Angle between cutting force component F_{xy} and the chip-thickness change component u_{xy}
β_{orth}	deg	Angle between the direction of the dynamic cutting force and the direction of chip-thickness change
$\gamma^2(\omega)$	—	Coherence function
ε	deg	Angle between the inner and outer chip-thickness change
κ	deg	Approach angle in a tool reference system
κ_M	deg	Approach angle in a machine co-ordinate system*
μ	%	Overlap factor
ρ	kg m^{-3}	Density
ϕ	deg	Phase angle
ϕ	deg	Tool angle in the xy plane (positive about the axis of rotation)
ϕ_x, ϕ_y, ϕ_z	deg	Angular deflections about axis x, y, z respectively
ω_n	s^{-1}	Natural gyro-frequency (undamped system)
ω_{dn}	s^{-1}	Natural gyro-frequency (damped system)
ω_R	s^{-1}	Resonance gyro-frequency (damped system)

*The angle κ_M is introduced to obtain a clear definition of the position of the cutting edge in a machine co-ordinate system. In contrast to the approach angle κ, as defined in DIN 6581, κ_M does not describe the angle between the cutting edge and the operating plane of the tool (the tool reference system); instead it represents the angle between the cutting edge and the plane perpendicular to the axis of rotation of the tool or workpiece measured in the direction of the positive z axis.

Indices/suffixes

Cal	Calculated
cr	Critical, limited
d	Direct
dyn	Dynamic
i	In
i, n, m	Running variables
max	Maximum
N	Nominal
neg	Negative
o	Out
St	Disturbance, interference
stat	Static
T	Translatory
TL	Tool
Try	Trial
WP	Workpiece
x, y, z	Relative to the co-ordinate directions

1

AIMS AND METHODS FOR DETERMINING THE VARIABLES

Work accuracy (Chapter 6), performance capabilities (Chapter 7) and the effect upon the environment of machine tools are dependent upon many characteristics (Fig. 1.1). The examination of these attributes may be organized to achieve two different aims and follow two different lines of approach, as shown in Fig. 1.2. On the one hand the aims are directed towards the establishment of any shortcomings of the machine with a view to improving its characteristics and on the other hand the intention is to arrive at an overall judgement when a machine is delivered to the user.

'Direct determination' of the machine characteristics requires the measurement of each individual attribute. This enables a precise identification of

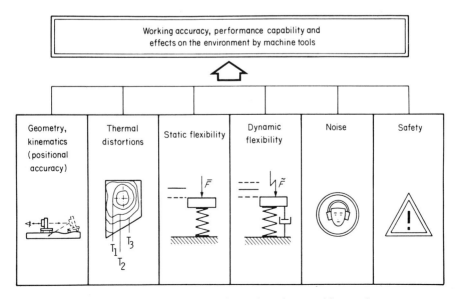

Fig. 1.1   Characteristics of metal-cutting machine tools

1

2

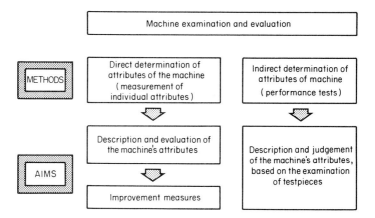

Fig. 1.2 Aims and methods of machine examination and evaluation

the causes of error and thus provides an objective basis for the introduction of the necessary steps to initiate improvement measures.

With respect to the delivery of the machine to the user, this method enables the individual characteristics to be ranked in order of their importance and permits them to be quantified according to their influence on the work accuracy and performance capabilities. Some serious difficulties are presented which, when considered alongside the need for considerably expensive measuring and testing equipment, may stand in the way of its application.

'Indirect determination' of the machine characteristics, i.e. the use of performance tests with associated examination of the finished testpiece, permits, in contrast to the previously mentioned method, only a limited determination of the causes which will allow a series of controlled improvement measures to be introduced. This is because the influence of the individual characteristics upon the work accuracy and performance capabilities cannot be separately identified. Only in rare cases can conclusions be reached in respect of the contribution of the individual characteristics to the overall result.

When describing and appraising machine characteristics based upon the results of performance tests, the objectivity of the method is made more complicated by the fact that, for example, the cutting process characteristics also influence the overall result. Thus due to tool wear or variations in work or tool materials or inconsistent sizes of work-load, conclusions are difficult to reach with respect to the actual machine characteristics. As these boundary cutting conditions are only partially known, such judgements on the performance capabilities when accepting the machine are sometimes subject to considerable unreliability or, as is often the practice, are only possible with the support of statistical information. Thus this method also has its high costs as a limiting factor.

The acceptance charts for machines (the statement from the manufacturer for the user with regard to the quality of the machine) should have the inspection criteria and the permissible deviations based upon international or national standards or acceptance recommendations from professional associations, whenever any of these are available. It is, of course, always possible for there to be special arrangements in that respect between the manufacturer and purchaser.

At the international level, the ISO (International Organization of Standardization) co-ordinates the standardization efforts of the member nations. At the national level in the Federal German Republic the standard association known as Normenausschuss Werkzeugmaschinen (NWM) (Machine Tool Standards Association) covers this work for machine tools. The procedures to be adopted are published in the form of DIN standards.[1,2] The initiative for such standards comes from professional associations: VDI (Verein Deutscher Ingenieure—Association of German Engineers), DGQ (Deutsche Gesellschaft für Qualität—German Association for Quality) or manufacturers associations (e.g. VDW, Verein Deutscher Werkzeugmaschinenhersteller—Association of German Machine Tool Manufacturers), who do this work partially themselves and co-operate with the NWM. These relationships are clarified in Fig. 1.3.

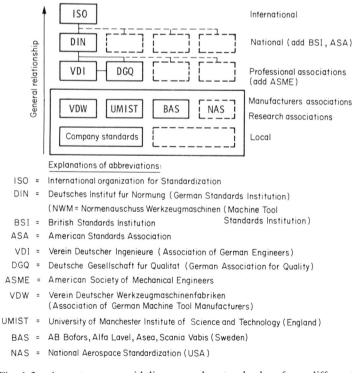

Explanations of abbreviations:

ISO = International organization for Standardization
DIN = Deutsches Institut fur Normung (German Standards Institution)
(NWM = Normenauschuss Werkzeugmaschinen (Machine Tool Standards Institution)
BSI = British Standards Institution
ASA = American Standards Association
VDI = Verein Deutscher Ingenieure (Association of German Engineers)
DGQ = Deutsche Gesellschaft fur Qualitat (German Association for Quality)
ASME = American Society of Mechanical Engineers
VDW = Verein Deutscher Werkzeugmaschinenfabriken (Association of German Machine Tool Manufacturers)
UMIST = University of Manchester Institute of Science and Technology (England)
BAS = AB Bofors, Alfa Lavel, Asea, Scania Vabis (Sweden)
NAS = National Aerospace Standardization (USA)

Fig. 1.3 Acceptance guidelines and standards for different validations

4

1. *Accuracy tests*
   1.1 Geometric accuracy by manufacture
   1.2 Influence of weight of work
   1.3 Influence of positional changes of moveable units
   1.4 Setting of dimensions by hand
   1.5 Consistency of feed
   1.6 Main motion
   1.7 Clamping
   1.8 Deformations due to heat
   1.9 Forced vibrations

2. *Tests for static stiffness*
   2.1 With respect to accuracy (only for certain machines)
   2.2 With respect to stability (only for particular parts of the machine, identified in a given case)

3. *Tests for limiting conditions to produce cutting stability*

4. *Tests for noise (volume, source)*

5. Tests of automatic control governing the relative motion between tool and work (also examined will be; resultant accuracy, limiting parameters for velocity, acceleration, forces and reliability. In addition tests are carried out for the design characteristics of the control system)
   5.1 Devices for limiting the feed movements
       5.1.1 System of stops
       5.1.2 Devices with limit switches
       5.1.3 Cam units
       5.1.4 Digital/numerical setting
   5.2 Devices to control path of movement
       5.2.1 Copying systems
       5.2.2 Numerical control
   5.3 Tests of devices which activate controls

6. Tests of mechanical and automated transportation, storage and loading devices

7. Tests of mechanized and automated devices for tool changing

8. Tests of clamping devices for work and tools

9. Tests of main drive (the drive includes all components which transmit power to the spindle or table)

10. Tests of feed-drive (the drive includes all components which transmit power to the table or tool holder)

11. Tests of guideways (rotary and linear motion)

12. Tests fof operating components for manual control

13. Tests of electrical installations

14. Tests of pneumatic systems

15. Tests of hydraulic systems

16. Tests of lubrication systems

17. Tests of coolant systems

18. Tests of suction units

19. *Tests of safety devices* (for protection of operator and machine)

20. Technical–economic evaluation
    20.1 Technological tests
    20.2 Technical–economic evaluation of the machine

21. Tests for rigidity and wear resistance

Fig. 1.4   Glossary of acceptance tests (Tlusty)

In the existing standards, acceptance standards are specified only for the geometric features,[1,2] positional accuracies, working accuracies,[3] noise emission[4] and safety. The individual chapters of this book deal with these standards and further developments in these fields. In addition the latest metrological techniques, the equipment necessary and the interpretation of its output are presented. In order to understand the effects of different criteria and the means available for their improvement, the technological and physical relationships are explained where necessary.

A broad classification of the multitude of acceptance tests which are possible is presented in Fig. 1.4. The choice of the tests and their permissible tolerances which are to be applied in a given instance are based upon the expected performance requirements. The general principle which should be followed is that 'a machine need only be as good as the range of products to be produced on it demands'.[5]

The fundamental procedure to be followed when determining the characteristics of a machine tool based upon the quality features laid down is shown in Fig. 1.5. A most important point is the definition of these features which will provide the criteria for appraising the characteristics of the machine under test. After the parameters have been measured with suitable measurement devices and the results have been statistically collated, they must be evaluated and interpreted. This is done by comparing the measured values with those which have been laid down to satisfy the performance expected from the machine. Moreover, it is necessary to clarify the relationship which exists between the parameter under consideration and the other machine characteristics.

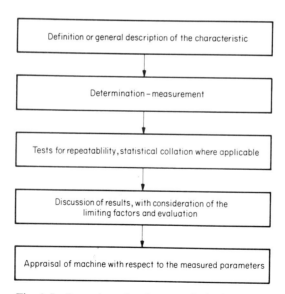

Fig. 1.5   Procedure for the appraisal of machine
characteristics

Periodic/systematic errors: Inherent in system
Reproducible for given boundary condition
Equal in value and direction at each measurement
Follows mathematical law; hence may be
    compensated

Examples: Errors in calibration of measuring system
Geometric errors in guides
Reversal error
Temperature (working)

Random errors: Cannot be reproduced even for given boundary
    conditions
Does not follow mathematical law but can be
    determined statistically

Examples: Friction effects
Bearing clearance gaps
Temperature (ambient)

Fig. 1.6 Periodic/systematic errors and random errors in machine-tool appraisal

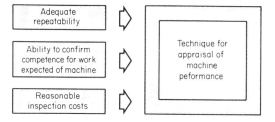

Fig. 1.7 Basic requirements of techniques for
the appraisal of machine performance

By applying statistical evaluation techniques, differentiation may be made between periodic or systematic errors on the one hand and random errors on the other. Figure 1.6 explains these different categories with the aid of examples.

The major requirements of a technique suitable for the appraisal of machine characteristics are shown in Fig. 1.7. In the first instance an adequate accuracy and repeatability of the test results must be available. The procedure should be sufficiently broad to enable tests for as many functions (working conditions, tool–workpiece configurations, etc.) as possible to be carried out. In order to ensure that the technique is economic and is not limited to labour-intensive inspections, the measurements must be able to be carried out at acceptable costs.

DIRECT MEASUREMENT AND APPRAISAL OF MACHINE CHARACTERISTICS

2

GEOMETRIC AND KINEMATIC BEHAVIOUR OF MACHINES

The accuracy of work produced on metal-cutting, forming and erosion machine tools is largely influenced by the following:

(a) deviations from the planned relative movement between the tool or tool carrier (e.g. spindle, quill) and the work or work support (e.g. table);
(b) condition (wear) and elastic deformations of the tool;
(c) elastic deformations of the work and clamping elements.

The first-mentioned factor is attributable to the machine; for this reason it will be examined in some detail in this volume.

The causes of deviations from the defined relative motions between the tool and work may be grouped into geometric and kinematic categories. Under the term 'geometric deviation' are grouped positional inaccuracies and errors in the shape of the machine components (tables, tool holders, guides, quills), whilst 'kinematic deviations' occur in co-ordinate movements, i.e. functional movements (thread cutting, guideway control). These errors are transmitted into the accuracy of the workpiece in varying amounts, depending on the particular production conditions which apply in a given case.

Both types of deviations are the result of production and assembly errors in the elements which are used in the machine construction, as well as their elastic deformations due to static, dynamic or thermal loading conditions as shown in Fig. 2.1.

2.1 Geometric deviations

When describing the expected deviations within the whole operating space of a machine the individual contributions of all moving axes must be considered.

10

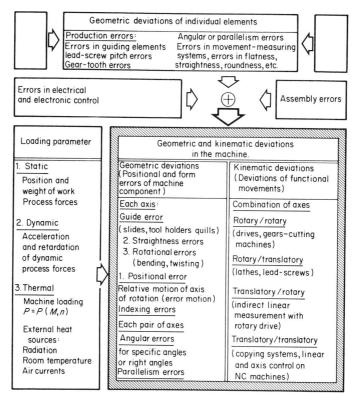

Fig. 2.1   Factors influencing geometric and kinematic deviations
in a machine

2.1.1   *General description of periodic or constant deflections and errors*

2.1.1.1   *Movement along one axis*

The geometric deflections of movement of a machine component such as a
slide, in every possible position along its axis of motion, can be referred to six
'datums', in analogy with the six degrees of freedom of a body in space.

If we consider a point P_0 in the plane of a machine table, then three
translatory deflections may be identified in the orthogonal axis directions, as
shown in Fig. 2.2, where:

$\delta x_T(x)$ = positional error, with deflection in the direction of feed (reference
axis)

$\delta y_T(x)$ = straightness error, with deflection perpendicular to the direction of
feed

$\delta z_T(x)$ = straightness error, with deflection perpendicular to the direction of
feed

In order to be able to determine the deflection from the theoretical position

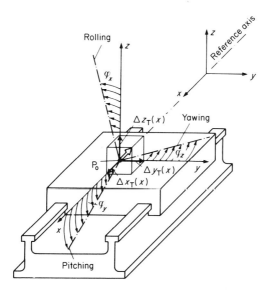

Fig. 2.2 Three translatory and three rotary deflections of a machine component with one axis of movement

of a particular point P_i with reference to the point P_0 within the operating space of the machine, as shown in Fig. 2.3, it is necessary to consider angular deviations ϕ_x, ϕ_y, ϕ_z in addition to the translatory deflections, as shown in Fig. 2.2 (compensation for the Abbé errors because the point P_i does not lie on the reference axis), where:

ϕ_x = angular error about the x axis (rolling movement; rotation about the feed axis)

ϕ_y = angular error about the y axis (pitching or tipping movement; rotation about one axis in the plane of the table perpendicular to the feed motion)

ϕ_z = angular error about the z axis (yawing or rotary movement; rotation normal to the plane of the table)

Assuming that the table itself does not distort, the above data may be used to define the deflections or errors of every point on the workpiece within the working space. As it may be assumed that angular errors ϕ_x, ϕ_y and ϕ_z are small, the sines of the angles may be substituted by the respective arcs and the cosines may be taken to be equal to 1. The deviation $\{a\}$ $(x_0 . x_{WP_i})$ at the component point P_i, with its co-ordinates $\{x\}_{WP_i}$, $\{y\}_{WP_i}$ and $\{z\}_{WP_i}$ when measured from the reference point P_0, is then given by:

$$\{a\}(x_0, X_{WP_i}) = \{a\}_T(x_0) + [A]_{\leqslant}(x_0)\{X_{WP_i}\} \qquad (2.1)$$

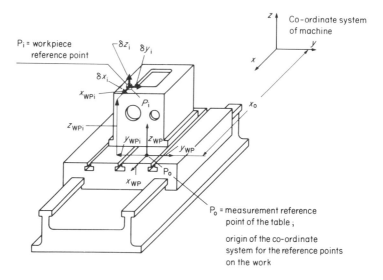

Fig. 2.3   Arrangement of co-ordinates for machine and workpiece

and

$$\left\{\begin{matrix} \delta x \\ \delta y \\ \delta z \end{matrix}\right\}(x_0, X_{WP_i}) = \left\{\begin{matrix} \delta x \\ \delta y \\ \delta z \end{matrix}\right\}_T(x_0) + \begin{bmatrix} 0 & -\phi_z & \phi_y \\ \phi_z & 0 & -\phi_x \\ -\phi_y & \phi_x & 0 \end{bmatrix}(x_0)\left\{\begin{matrix} x_{WP_i} \\ y_{WP_i} \\ z_{WP_i} \end{matrix}\right\} \qquad (2.2)$$

where
x_0 = the machine co-ordinate, in this case in the direction of feed

X_{WP_i} = the component co-ordinates measured from the reference point P_0

$\{a\}_T(x_0)$ = the deflection vector for the three translatory components measured from the reference point P_0 in the slide position x_0. In addition to this deflection, the effect of the rotary movements of the slide must be considered for all points P_i outside the reference point, which can be expressed by the second term of the equation, where:

$[A]_{\triangleleft}(x_0)$ = the matrix for the angular displacement of the table in its position x_0, the machine co-ordinate and

$\{X_{WP_i}\}$ = the workpiece co-ordinates of the point on the work P_i which have their origin at the reference point P_0 on the table.

The metrological procedures for the determination of the individual components of the errors are described in detail in section 2.1.5.

2.1.1.2 Movement along several axes

In addition to the above-mentioned deflections, which may be determined for every axis of movement in the manner indicated, when there are multi-axis movements the angular deflections of the axes with respect to each other must also be taken into account. The additional error caused by the angular deflections of the axes may be determined from the previously used definition by the matrix for the angular axis deflections $[A]_{\triangleleft 0}$ and the work-space co-ordinates x, y, z. As shown on the left of Fig. 2.4 the angles ϕ_{xy}, ϕ_{xz} indicate the angular axis errors of the x axis and correspondingly the angles ϕ_{yz}, ϕ_{yz} or ϕ_{zx}, ϕ_{zy} the angular axis errors of the y or z axes respectively about the axes of the reference co-ordinate system.

The deflections of a point in the work-space of a machine due to such angular axis errors are directly proportional to the distance travelled along the co-ordinate axes of the machine system:

$$\{a\}_{\triangleleft 0}(X) = [A]_{\triangleleft 0}[X_0] \tag{2.3}$$

where

$$[A]_{\triangleleft 0} = \text{the matrix for the measured angular deflections of the axes with respect to each other}$$
$$\{x_0\} = \text{the positional vector of the guided machine component within the machine co-ordinate system } x, y, z$$

Thus:

$$\begin{Bmatrix} \delta x \\ \delta y \\ \delta z \end{Bmatrix}_{\triangleleft 0} (X) = \begin{bmatrix} 0 & -\phi_{yz} & \phi_{zy} \\ \phi_{xz} & 0 & -\phi_{zx} \\ -\phi_{xy} & \phi_{yx} & 0 \end{bmatrix} \begin{Bmatrix} x_0 \\ y_0 \\ z_0 \end{Bmatrix} \tag{2.4}$$

If one axis, e.g. the x axis, is chosen as the datum axis and the x, y plane (table

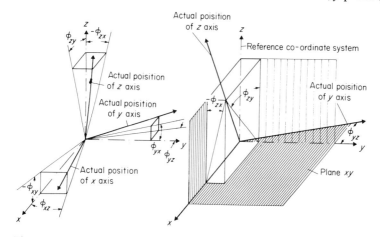

Fig. 2.4 Angular deflections of the axes of movement with respect to each other

14

plane) as the datum plane (Fig. 2.4, right) then the deflections are independent of the x co-ordinate and the deflections in the z direction closely approximate zero. This simplifies the matrix of the angular axis deflections to be measured and we have:

$$\begin{Bmatrix} \delta x \\ \delta y \\ \delta z \end{Bmatrix}_{<0} (X) = \begin{bmatrix} 0 & -\phi_{yz} & \phi_{zy} \\ 0 & 0 & -\phi_{zx} \\ 0 & 0 & 0 \end{bmatrix} \begin{Bmatrix} x_0 \\ y_0 \\ z_0 \end{Bmatrix} \qquad (2.5)$$

In the case of a three-axis movement, the total deflection of a point within the working space of a machine is given by:

$$\{a\}(X_0, X_{WP_i}) = \{a\}(x_0, X_{WP_i}) + \{a\}(y_0, X_{WP_i}) + \{a\}(z_0, X_{WP_i})$$
$$+ \{a\}_{<0}(X_0) \qquad (2.6)$$

or $\quad \{a\}(X_0, X_{WP_i}) = \{a\}_T(x_0) + \{a\}_T(y_0) + \{a\}_T(z_0) + ([A]_<(x_0)$
$$+ [A]_<(y_0) + [A]_<(z_0))\{X_{WP_i}\} + [A]_{<0}\{X_0\} \qquad (2.7)$$

The following must be considered to determine the appropriate signs in the equations. As the relative displacement of the tool with respect to the work is the important quantity the difference between the deflections of work support and that of the tool holder must be established. To this end the individual deflections of the appropriate axes which affect the motion of the work support on the one hand and those which influence the tool holder on the other must be added. Hence, for the case depicted in Fig. 2.5 as an example, the deflections of the axes x and z (work support) must be added, the errors in the axes y and w (tool holder) are similarly summed and the difference between these totals is then established. Thus:

$$\{a\} = (\{a\}(x_0, X_{WP_i}) + \{a\}(z_0, X_{WP_i})) - (\{a\}(y_0, X_{TL_i}) + \{a\}(w_0, X_{TL_i})) \quad (2.8)$$

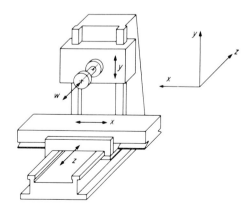

Fig. 2.5   Example of a machine tool with
multi-axis movements

The geometric deflections of a point on the workpiece on a three-co-ordinate machine are derived according to equation (2.7) from three times three trans-latory deflections plus three times three angular deflections and three angular axis deflections, i.e. a total of twenty-one individual deflections.

Limitation of results. This already somewhat complicated definition of the geometric relationships may be simplified by assuming that the deformations of the individual axes as a result of the dead-weights of the separate machine components are independent of each other. This, however, is generally not the case, as the various elastic deformations of the machine are governed by the position within the work-space of the measured deflections of any axis, produced by the weight of the machine components. Consequently the angular and translatory errors must be established as far as is possible for all the axes within the whole of the work-space. This is done by taking readings along the axes with the machine components in different positions. Equation (2.7) is thus modified to:

$$\{a\}(X_0, X_{WP_i}) = \{a\}(x_0(y_0, z_0), X_{WP_i}) + \{a\}(y_0(x_0, z_0), X_{WP_i})$$
$$+ \{a\}(z_0(x_0, y_0), X_{WP_i}) + [A]_{<0}\{X_0\} \qquad (2.9)$$

This resultant relationship is of course only valid for a specific loading condition, as may be noted from Fig. 2.1 (the position and weight of the workpiece, machine temperature, etc.). To enable a comprehensive evalua-tion of the geometric behaviour of the machine to be made a number of readings have to be obtained for various loading conditions.

In order to keep the costs of the evaluation within reason, it is necessary to define representative positions of the machine components for the measure-ments of the axial deflections according to the type and design of the machine. To this end, a wide range of standards needs to be made available in the future.

2.1.2 General description of the random contributions to the deflections

In addition to the systematic deflections described so far, which under given conditions may be reproduced and, providing the expenditure is warranted, may be corrected and compensated, random errors may also be noted. Such random or non-systematic deflections may vary from test to test in both magnitude and sign, and hence statistical conventions are applied in order to provide their description.

Figure 2.6 gives an example of the results which could be obtained for one deflection parameter over the length of travel along one axis. Next to the mean-value curves, which are established for both directions of travel from several readings (at least five) under the same boundary conditions, the scat-ter band ($2s$ or $3s$) is shown. (Derivations of these values from the individual readings are explained in section 2.1.5.2.) The difference between the mean

16

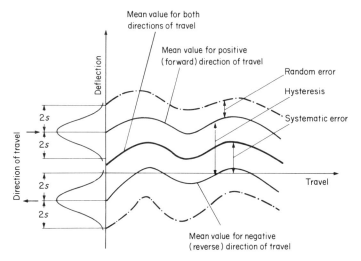

Fig. 2.6   Presentation of the errors for one parameter over the
length of travel along one axis

values obtained for each direction of travel is the hysteresis or backlash which
is governed by the play in the transmission components and the friction
forces.

The measurements of the deflections for individual parameters do not produce a single meaningful curve for the axis of movement but produce a more
or less wide scatter band due to the random deflections further widened by
the hysteresis.

2.1.3   Determination of workpiece dimensional errors due to the geometric machine deflections

The geometric deflections of the machine components from their nominal
position as described above are responsible for workpiece errors which
develop during the production process on the machine in question. As the
work measurements, e.g. length measurements, are subject to the travel of
the appropriate table, tool holder or spindle over that particular distance, the
difference between the machine deflections in the two extreme positions of
this movement is of importance.

The workpiece errors are given by:

$$\{a\} = \{a\}\ (X_{01}, X_{\mathrm{WP_i}}) - \{a\}\ (X_{02}, X_{\mathrm{WP_i}}) \tag{2.10}$$

where X_{01} and X_{02} are the co-ordinates of the machine components for the
corresponding work dimension. When the random deflections are taken into
account, we obtain, as clarified in Fig. 2.7, a scatter band of width $2s$ for each
position in addition to the systematic deflections of the machine in the positions ① and ② . Within that band, 95% of all values are represented. The
statistical certainty that the work size will be within the limits $l_{\min}$ and $l_{\max}$ is,
according to statistical theory, approximately 99%.

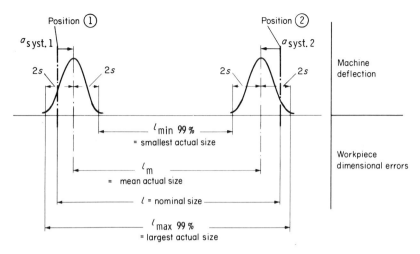

Fig. 2.7   Relationship between geometric machine errors and workpiece dimensional errors

2.1.4   Linear treatment of deflections

Attempts have been made to simplify the treatment of geometric deviations by a linear approach.[6,7] If the three translatory deflections including their random portions are to be obtained for the whole of the working space of a machine by measurement, then the multitude of readings that will be necessary requires a simplified presentation to be developed.

One possible approach is to identify the greatest difference in deflection occurring between defined positions, so that all geometric dimensional errors will be certain to be with these limits. Experience of a large number of cases has, however, indicated that systematic deflections frequently occur—at least within certain ranges—which are proportional to the distance along which the machine units are moved. Consequently, the criteria defined above (maximum deflections in the whole of the working space) are too severe as an indication of the quality of the machine, particularly with respect to smaller-sized workpieces.

By describing the machine deflections in relation to the length of travel, the most suitable machine may be selected for given production conditions and requirements (tolerance, size of workpiece). Moreover, the opportunity is provided to classify the geometric quality of the machine in accordance with ISO definitions as a first approximation, because for given qualities the tolerances of a workpiece are similarly a function of its size:

$$i = 0.001l + 0.45 \sqrt[3]{l} \tag{2.11}$$

where l is in millimetres and i is in micrometres for quality 1.

The deflections plotted in Fig. 2.8 with respect to the distance travelled are interpreted against an imaginary theoretical template. This template must be

18

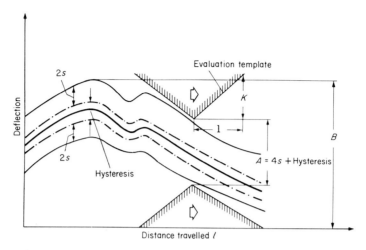

Fig. 2.8 Evaluation template for a particular group of measured deflections

chosen in such a way which will permit it to be moved parallel to the pattern of curves whilst remaining close to the curve. The narrowest gap on the template determines the distance A; A is mainly constituted from the random portion of the deflections and is evaluated as: $A \geq 4s$ + hysteresis, measured at the most unfavourable position of the curve. The slope K of the template is determined by the uncorrected systematic error in relation to the length travelled. The deflection for a given position of the machine unit can therefore be stated in relation to the distance travelled l as:

$$U = A + Kl \leq B \qquad (2.12)$$

where: [6,7]

U = deflection (uncertainty)
A, K = template dimensions (K must always be positive)
l = distance travelled
B = the maximum error noted

Figure 2.9 depicts this relationship diagrammatically. The limiting value B is of importance since in general an increase in deflection does not occur over the whole of the length of movement.

The constants A and K must be of such a magnitude that equation (2.12) envelopes the curve in every case. Theoretically, this condition may be satisfied by an infinite number of combinations of these values. A more practical procedure for the establishment of values A and K is shown in Fig. 2.10. In this method, the spread of the curves is approximated by a series of rectangles which are evenly divided (in this case ten divisions) over the whole of the path of motion. The longest rectangle (in this case between 9 and 10) is designated to be of height A. This is then placed at the point where the systematic deflections are at a maximum (or minimum) (in this case between 4 and 5).

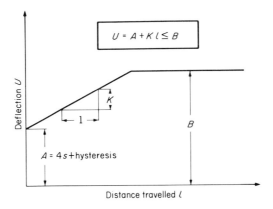

Fig. 2.9 Linearized curve of a deflection pattern

This then enables the slope K for the template to be established for the side being considered.

By application of equation (2.12) the upper limits of the dimensional errors of the work may be determined. As explained in section 2.1.3 the dimensional errors of a workpiece are a result of the differences between the relative deflections of the tool-holding and work-mounting components of the machine, whereby both the systematic and random components must be taken into account. As in equation (2.12) the systematic deflections are presented in absolute terms, i.e. without signs. The workpiece dimensional deviations from their theoretical value are therefore given by:

$$U_{WP} = \pm (A + Kl) \tag{2.13}$$

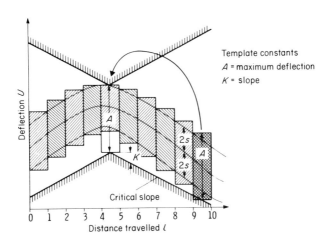

Fig. 2.10 Determination of the constants for an evaluation template from a given set of measured readings

20

The unidirectional approach applied to the examples used so far may be extended for problems dealing with a plane or a space. In the case of a three-dimensional space, the following system of equations is valid:

$$\begin{Bmatrix} U_x \\ U_y \\ U_z \end{Bmatrix} = \begin{Bmatrix} A_x \\ A_y \\ A_z \end{Bmatrix} + \begin{bmatrix} K_{xx} & K_{xy} & K_{xz} \\ K_{yx} & K_{yy} & K_{yz} \\ K_{zx} & K_{zy} & K_{zz} \end{bmatrix} \begin{Bmatrix} x \\ y \\ z \end{Bmatrix} \leqslant \begin{Bmatrix} B_x \\ B_y \\ B_z \end{Bmatrix} \qquad (2.14)$$

Figure 2.11 shows a three-dimensional presentation of a linearized deflection pattern in direction x for a movement in a xy plane. This obviously simplified presentation of the deflections is obtained, however, at an increased inspection cost. Determination of the angular errors of the machine components may be eliminated because the linear deflections will be measured in the whole of the working space. Apart from the disadvantage of high inspection costs, there are at present no satisfactory techniques available which may be used accurately and meaningfully to determine the constants in the equations (A_{ij} and K_{ij}). This is particularly the case for multi-axial problems.

With the ever-increasing use of numerically controlled machine tools on which components have to be produced to close dimensional tolerances, as well as the growing application of three-dimensional measuring machines in production, the need for the development of an objective method for geometric acceptance tests and a clear description of the deflections in the working space is becoming more and more urgent. The future will show which of the methods presented here will enable a meaningful definition of the deflections to be produced.

The development of measuring devices is closely related to the determination of deflections. The individual linear and rotary deflections can today be relatively simply determined and recorded with the aid of modern metrological instruments.

The following sections deal with the currently used metrological techniques and equipment for geometric measurement.

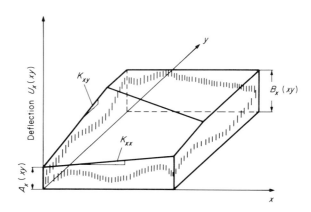

Fig. 2.11   Linearized deflection U_x within the xy plane

2.1.5 Metrological techniques to determine geometric characteristics of machines

The examination of the geometric characteristics of machine tools comprises tests for size, shape and position of the machine components, insofar as these affect the machine's capabilities for accuracy. When these measurements are carried out the machine is not normally subjected to cutting forces or the dead-weight of workpieces. It is, however, reasonable to assume that these latter influences will affect the static, dynamic and thermal machine characteristics, as well as its geometric behaviour. Hence it is sensible and desirable to conduct geometric machine appraisals under defined and representative loading conditions.

The parameters which must be examined may be established from draft publications by the German Standards Institution (DIN) which have been expanded by the Standards Association for Machine Tools (NWM) (DIN 8601 ff.).[1,2] These tests are carried out at the machine manufacturers' plants with appropriate measuring devices.

The standards specifications make recommendations for the most suitable metrological instruments and methods of measurement. These instruments are generally to be found among the basic equipment of every workshop. However, the specifications will permit alternative equipment which may be available to be used, providing that their degree of accuracy corresponds at

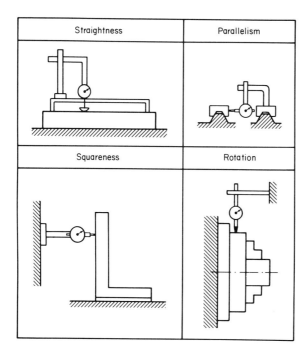

Fig. 2.12   Criteria for geometric alignment testing

least to that of the recommended instruments. This latitude, which allows the application of measuring techniques other than those stated in the standard, is often used when the machines or machine components which are to be inspected are very large or when the accuracy required from the machine is exceptionally high.

For such cases, a number of techniques based upon modern technologies such as laser beams, optoelectronics, as well as computer data processing and evaluation, have been developed and applied in practice during recent years. These enable a large number of basic measurements of the geometric (and kinematic) machine characteristics to be taken. In addition, the opportunity is provided to automate the inspection procedure to a large extent, as well as enabling the measured values to be recorded and evaluated directly.

Figure 2.12 shows some fundamental criteria of geometric machine-tool alignment evaluation and the principal measuring set-ups using conventional metrological instruments (straight edge, precision square, dial gauge, etc.). In Fig. 2.13 an extract from the geometric acceptance conditions for lathes (DIN 8605) is reproduced. The objectives of the tests (features to be examined), a diagrammatic presentation of the set-up, the equipment required and instructions for the performance of the tests for the individual stages of examination are all arranged in tabulated form. In addition the permissible errors are stated and a further column provided in which the readings obtained by the measurements may be recorded.

In the following paragraphs some fundamental geometric tests are

	Test objectives	Diagram	Test equipment	Test procedure	Errors	
					Permissible	Measured
4	(a) Axial spindle float	b, a, F	Precision dial gauge (accuracy grade 1 to DIN 879)	Precision dial gauge (a) On centre of revolution (b) On spindle face	a 0.005 b 0.01 (Inclusive of axial spindle float)	
	(b) True running accuracy of spindle face		Attachments as required	Turn spindle slowly. If there is play in the thrust bearings, apply a constant force F The value of F to be specified by the manufacture. 5.6.2 5.6.3		
5	Concentricity of centering cone of spindle	F	Precision dial gauge (accuracy grade 1 to DIN 879)	Set dial gauge perpendicular to circumference of cone Turn spindle slowly If there is play in the thrust bearings apply constant force F. The value of F to be specified by manufacturer. 5.6.1.2.2	0.007	

Fig. 2.13 'Acceptance tests for lathes', extract from DIN draft 8605

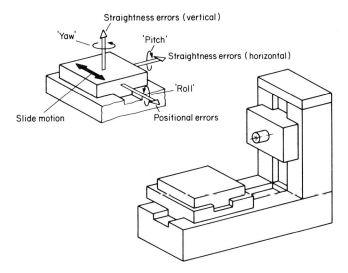

Fig. 2.14   Alignment errors of a machine-tool slide

described using the horizontal boring and milling machine shown in Fig. 2.14 as an example. The deflections which are determined by these measurements come, in accordance with the classifications given in Fig. 2.1 for geometric deviations, into the category of deflections of individual machine axes as well as that of deflections between several axes.

Geometric measurement of individual machine axes. Any one of the three slides of the horizontal boring and milling machine shown in Fig. 2.14 may be subject to the alignment errors shown in the upper left of the diagram. We are concerned here with translatory deflections in the direction of the three co-ordinate axes (positional error and two straightness errors) and with rotary deflections about these axes (three rotational errors). In addition the machine spindle may have rotation errors.

2.1.5.1   Measurement of straightness errors

Measurement with test wire and microscope. A steel wire of diameter 0.1 mm is fitted under tension parallel to the surface which is to be examined, as shown in Fig. 2.15. If the horizontal straightness of the line MN is to be measured, then a microscope may be mounted perpendicular to the guiding surface which will enable the errors in relation to the test wire in the measurement plane to be determined.

Measurement with a sighting telescope. A sighting telescope with built-in cross wires is arranged with its optical axis parallel to the surface under examination. Within the path of the light rays a sight graticule is fitted which

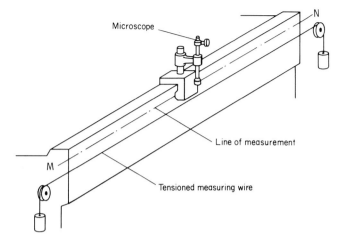

Fig. 2.15   Measurement of the flatness errors of a surface with a
measuring wire and microscope (Leinweber)

is guided along the line being inspected by the slide under test. Any deviation
of the graticule from a straight path may then be observed on the cross wires
in the telescope. The basic measurement set-up is illustrated in Fig. 2.16.

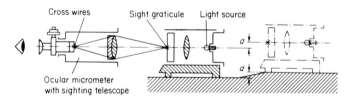

Fig. 2.16   Straightness measurement with a sighting telescope
(DIN 8601)

Measurement with a four-quadrant photodiode. In this method a focused light
beam is used as a reference for straightness measurement. The set-up is
shown in Fig. 2.17. A light beam (e.g. laser beam) is arranged parallel to the
path being examined (line of movement of the machine-tool table). This
beam falls on to a photodiode which registers the intensity of the incident
light on each of four quadrants separately. To obtain readings of straightness,
the photodiode is moved with the table along the path being inspected.
Movements in the plane perpendicular to the direction of travel produce
variations of light intensity in the four quadrants of the photodiode. By
suitable evaluations of the signals obtained from the quadrants the deviations
from a straight-line motion may be determined in both the vertical and
horizontal directions (see Fig. 2.18).

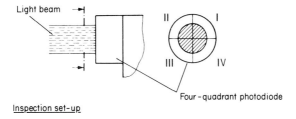

Inspection set-up

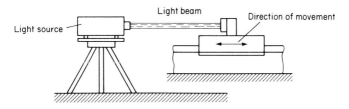

Fig. 2.17 Straightness measurement with a four-quadrant photodiode

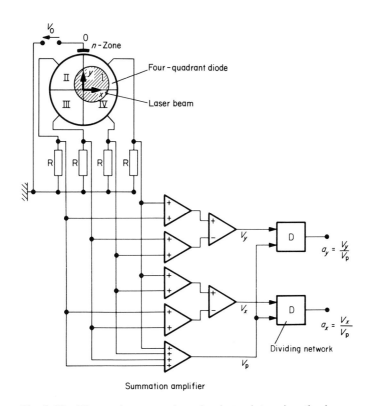

Fig. 2.18 Electronic summation circuit to determine the beam position with a four-quadrant diode

26

2.1.5.2 Measurement of positional errors

On numerically controlled machine tools, the positional accuracy of the feed units is of particular importance for the accuracy of the machine's output. Positional errors may be defined as the magnitude of the difference between the theoretical position and the actual position. This depends upon a number of factors, such as the resolution and accuracy of the linear measuring systems, elastic deformation of the drive components, the inertia forces when braking, friction and stick-slip effects in the guideways, table or slide movements resulting from clamping after positioning, etc.

The positional accuracy is further affected by the performance of the control system on NC machines and by the non-predictable errors of the operator on manually operated machines.

Measurement with a laser interferometer. In order to help in the understanding of this method the theory of linear measurement by laser interferometry is first explained. It is based upon the principles of the Michelson interferometer shown in Fig. 2.19. A light beam of a single wavelength and phase (mono-chromatic and coherent) impinges upon a beam-splitter S where it is divided into a reference and a measurement beam. Both beams are reflected by the mirrors M1 and M2, and interference occurs at these points. If the mirror M2 moves in the direction of the light beam, the photoelectric cell will detect reductions in light intensity, the wavelength of which corresponds to

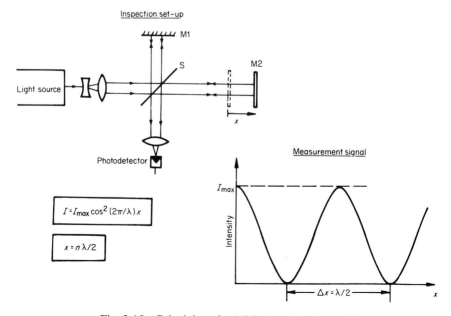

Fig. 2.19   Principles of a Michelson interferometer

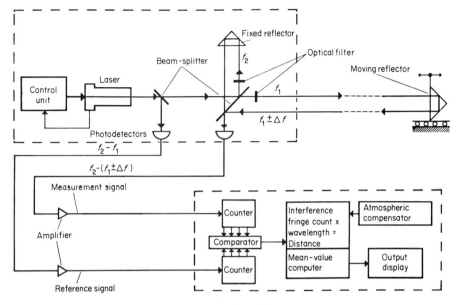

Fig. 2.20   Laser interferometer

the length of half of one wavelength of the light being used. From this signal impulses are produced and measured.

The accuracy of the laser interferometer technique is largely dependent upon the stability of the laser light wavelength. This is turn is influenced by environmental factors (air pressure, temperature, humidity, carbon dioxide content), as well as the operating conditions of the laser itself (warming-up phase, etc.). Consequently, the environmental conditions must be constantly monitored during the inspection procedure and the measured values corrected in accordance with any departure from standard values.

A number of possibilities exist to improve the stability of the laser beam's wavelength, e.g. the use of a double-frequency laser. This avoids long warming-up periods due to the automatic wavelength stabilization which is made possible. The laser is operational immediately after switching on. Moreover, the double-frequency laser provides the opportunity to determine in a simple way the direction of movement of the objective, which is a prerequisite for measuring dynamic movements. The functional principles are explained in Fig. 2.20. Two light sources with differing frequencies f_1 and f_2 are produced by the laser. At the first beam-splitter a part of the light from the laser is diverted to form the reference signal, which is produced by superimposing the two light sources to result in a surge of intensity with a frequency of $f = f_2 - f_1$. The remaining light from the laser reaches a second beam-splitter. Here an optical filter transmits the beam with frequency f_2 to a fixed reflector and the light with frequency f_1 is directed to another reflector which is attached to the objective being observed. The reflected beams produce interference fringes on the second beam-splitter. The fringe spacings obtained at the stationary measurement objective are equal to those of the reference signal. If the objective moves, then the reflected light beam is subject to a frequency variation which will be higher or lower depending upon the direction of movement. The fringe spacings also alter in accord. By obtaining the difference between the reference and the observed fringe spacings, the magnitude and direction of movement of the measurement objective may be determined.

Further information on the application of laser techniques for machine alignment and acceptance testing is contained in the appropriate reference literature.[8,9]

These methods permit measurement of translatory movements to be made to a resolution of 0.158 μm (helium–neon laser) with an accuracy of ± 1 μm m^{-1} at speeds up to 18 m min^{-1}.

In order to examine positional accuracies (positional errors) the slides are moved repeatedly to a number of reference positions for each axis from both directions of travel, where their actual positions are determined, e.g. with the aid of laser interferometers. Figure 2.21 shows the set-up for the examination of the z axis (the direction of axis of the spindle rotation) on a machining centre. Twenty reference positions are approached five times from each direction, producing two hundred readings.

Fig. 2.21   Inspection of positional accuracy of a NC machining centre with the aid of a laser interferometer

Hence the measurements and evaluation systems need to follow a routine which will simultaneously provide a comprehensive and objective inspection procedure. The machine acceptance tests are standardized in the guidelines VDI/DGQ 3441 to 3445.[3]

To describe the positional accuracy for any one reference point, the following data must be obtained (Fig. 2.22):

1. Systematic errors

$$\bar{\bar{x}}_i = \frac{\bar{x}_i\!\uparrow + \bar{x}_i\!\downarrow}{2}$$

$\bar{x}_i\!\uparrow$ — mean value for positive direction of approach
$\bar{x}_i\!\downarrow$ = mean value for negative direction of approach

2. Reversal errors (backlash)

$$U_i = |\bar{x}_i\!\downarrow - \bar{x}_i\!\uparrow|$$

3. Positional scatter width

$$P_{S_i} = 6\bar{s}_i \quad \text{where} \quad \bar{s}_i = \frac{s_i\!\uparrow + s_i\!\downarrow}{2}$$

s = standard deviation

$$s_i\!\uparrow = \sqrt{\left[\frac{1}{n-1}\sum_{j=1}^{n}(x_{ji}\!\uparrow - \bar{x}_i\!\uparrow)^2\right]}$$

(frequently designated σ)

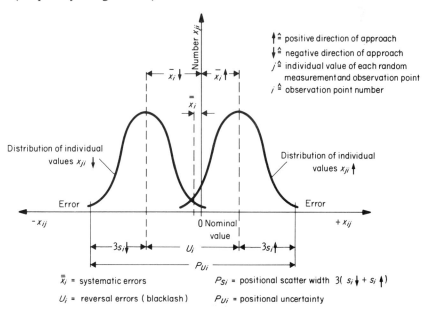

$\bar{\bar{x}}_i$ = systematic errors
U_i = reversal errors (blacklash)

P_{S_i} = positional scatter width $3(s_i\!\downarrow + s_i\!\uparrow)$
P_{U_i} = positional uncertainty

Fig. 2.22 Distribution of observed values, statistical data (source VDI 3254)

For a random sample size of $n < 25$ the standard deviation is calculated in accordance with the 'span-width' method (see guideline VDI 3254, sheet 1).

4. Positional error

$$P_{U_i} = U_i + P_{S_i}$$

Figure 2.23 is a reproduction of a characteristics chart for the positional accuracy of one axis of a milling machine which was obtained as a result of a systematic inspection evaluation. From this method of presentation, faults in the measurement system or feed mechanism may also be detected, e.g.:

(a) large amount of backlash in position 3 due to a fault in the feed shaft;
(b) steadily increasing error between positions 15 and 20 due to a lead-screw pitch error.

If the positional accuracies of a multi-axis machine for the whole of the working space were to be obtained, then the inspection costs would be of considerable magnitude. Standardization efforts are therefore directed towards development of measurement and evaluation procedures which provide a reliable picture of positional accuracy with acceptable expenditure.

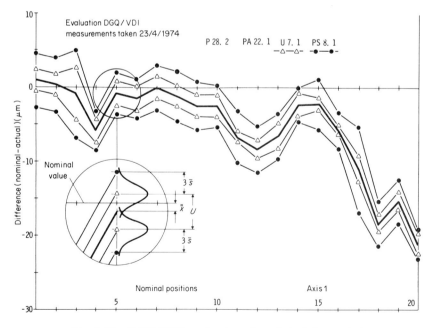

Fig. 2.23   DGQ/VDI evaluation of statistical data axis 1

2.1.5.3 Measurement of angular and rotary errors

Measurement with an auto-collimator telescope. A microscope and a telescope are arranged as shown in Fig. 2.24, coaxial and parallel to the plane of measurement. Every rotary movement by the moving mirror situated in the plane of measurement about its own axis results in a displacement of the cross-wire image in the plane of the light filament. This displacement is measured with the aid of an ocular-micrometer. By integrating the measured angle with respect to the distance travelled, the straightness of the surface can be determined.

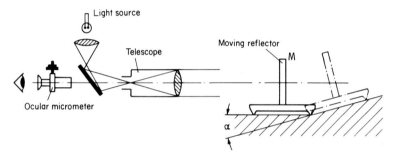

Fig. 2.24 Rotary measurements with an auto-collimator telescope (DIN 8601)

Measurement with a laser interferometer. With the use of suitable optical devices, angular movements may be measured by the combination of two linear measurements. Figure 2.25 shows the basic inspection set-up. Unlike the diagram shown in Fig. 2.20 the laser beam f_2 is not directed to a fixed reflector but to a second reflector which is also fixed to the measuring

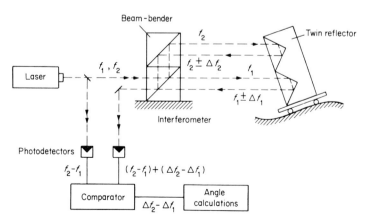

Fig. 2.25 Measurement of angular motion with a laser interferometer

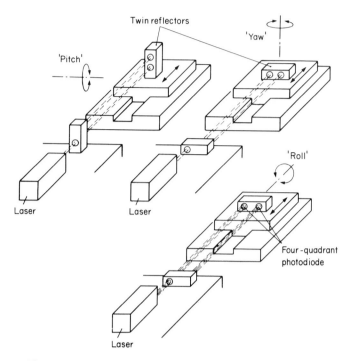

Fig. 2.26   Measurement of 'pitch', 'yaw' and 'roll' motions

objective. Both reflectors are at differing distances to the axis of the rotary movement. For this reason the two beams are subject to different frequency modulations f_1 and f_2 when the mirror on the object being measured is tilted. The change in the fringe spacings at the interferometer output is equal to the difference $\Delta f_2 - \Delta f_1$. From this and from the geometric arrangement of the two reflectors, the angular motion of the component under test may be calculated.

The arrangement of the inspection units to determine the 'pitch' and 'yaw' motions of a machine-tool slide is shown in the two upper sketches in Fig. 2.26. The lower diagram shows the set-up for determination of the 'roll' motion. In this case four-quadrant photodiodes are used instead of twin reflectors (see section 2.1.5.1). The angular movement is determined from the differing intensity variations of the two diodes.

2.1.5.4   Measurement of the relative motion of the axis of rotation (radial motion errors)

Consider a rotating shaft (spindle), the axis of which is perpendicular to a reference plane. If there is an error in the circular path, then the intersection of that axis with the reference plane will describe a curve. A detailed examination of such errors about the angle of rotation of the shaft is very complex. Hence only those errors which are relevant to the particular application are normally recorded.

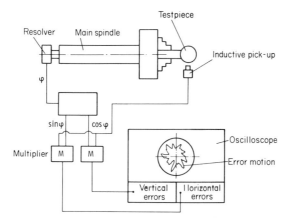

Fig. 2.27 Measurement of the relative motion of the axis of rotation (radial motion error) of a lathe spindle (Vanherck)

Figure 2.27 shows one possible method for measuring and simultaneously recording the relative motion of the axis of rotation, also known as the 'radial motion error'.[10] At the rear of the spindle the angle of rotation is measured by a resolver. The circuitry connected to it produces an output which is in terms of the sine and cosine of this angle. A spherical testpiece held in a chuck is used to measure the radial motion error with the use of an inductive pick-up; both angular functions are modulated and error signals are then fed to the horizontal and vertical inputs of an x/y recorder or an oscilloscope.

The eccentricity is measured only in the direction of the inductive pick-up, due to the particular method of setting. When such readings are taken on lathes, this direction is in line with the turning tool because this is the only direction in which any error will have an effect on the work accuracy.

2.1.5.5 Geometric measurement between pairs of machine axes

Squareness measurement between axes of motion. The determination of the squareness accuracy of two movements in respect to each other is shown in Fig. 2.28, using the procedure adopted on a plano-milling machine as an example. Firstly, the straightness of the table movement when moved in the z direction ($\delta y(z)$) is measured. For this purpose a four-quadrant photodiode (see section 2.1.5.1) is fixed to the table. The readings taken are recorded on the lower curves $\delta y(z)$ of the graph, shown in the bottom half of the diagram. The mean straight line produces an angle θ with the abscissa. This line indicates an error in parallelism between the light beam and the table guides, which is caused by inaccuracies in manufacture and is never completely avoidable.

At a second stage in the inspection procedure, a beam-bender (pentaprism) is placed upon the stationary table and positoned into the light beam path, so

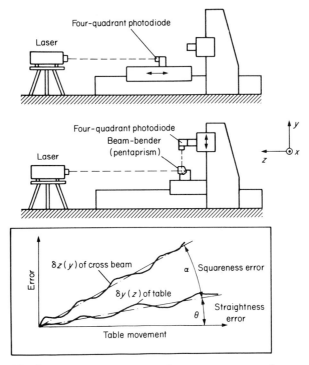

Fig. 2.28   Squareness measurement between two axes of movement

that the latter is diverted through 90° in the direction of the tool-holder slide movement y. The four-quadrant photodiode is fitted on to the tool holder. The straightness of travel $\delta z(y)$ of the tool-slide movement is now measured. The readings are plotted on the graph in the lower part of Fig. 2.28, resulting in the curve $\delta z(y)$. The mean straight lines of the two curves $\delta z(y)$ and $\delta y(z)$ produce an angle α between each other, which represents the error in squareness between the two machine movements.

Measurement of parallelism between axis of motion. Figure 2.29 shows a diagram of a suitable set-up for determining the errors in parallelism between two axes of machine movement. The light beam from the light source is arranged at right angles to the guideways and diverted to the level of the slide under test by means of a beam-bender. The beam falls on to a four-quadrant photodiode which is fitted to the work table. The table is then moved along the whole length of its travel in the z direction. This enables the straightness of movement in the x direction $\delta x(z)$ to be determined.

The same procedure is then adopted for the spindle movement of the boring and milling machine. To facilitate this test, the beam-bender is moved to the level of the spindle (chain-dotted position). The four-quadrant photodiode is fitted to the tool holder.

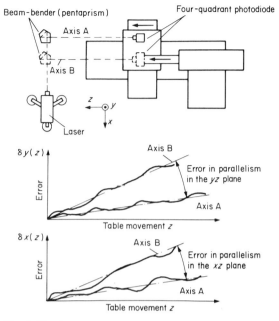

Fig. 2.29   Measurement of parallelism between axes
of movement

To determine the errors $\delta y(z)$, the laser, beam-bender and photodiode must be placed in the yz plane. The angle between the mean straight lines of the curves corresponds to the errors in parallelism in the xy plane and respectively in the yz plane of the two axes of movement (Fig. 2.29).

2.2   Kinematic errors

2.2.1   *General description*

The examination of the kinematic behaviour of machine tools is concerned with the relative motion errors of several moving machine components which are required to move in accordance with precise functional requirements. The tests are applied to those movements which influence the work accuracy of the machine. Although such measurements may be carried out in both the loaded and unloaded condition of the machine, the metrological difficulties encountered frequently require that the tests are undertaken in the no-load state. However, examinations which are made while the machine is subjected to cutting forces and inertia forces of the workpiece are of greater value.

The co-ordination of rotary motions with respect to each other is of particular importance for machine tools (e.g. gear-cutting machines) as are the rotary/translatory relationships (e.g. screw-cutting) and translatory/translatory relative movements (e.g. two-axis NC machines).

2.2.2 Inspection techniques to determine kinematic machine characteristics

To determine the kinematic machine errors, the movements must be measured with high resolution and accuracy. For this reason, linear movements are frequently measured with a laser interferometer. The principles of laser interferometry for linear measurement have already been dealt with in section 2.1.5.2. For the examination of rotary movements, incremental angle encoders (optical resolvers) may, for example, be employed.

Figure 2.30 shows the principle of operation of an optical resolver in diagrammatic form. A glass disc with radial gratings is rigidly connected to the rotating drive shaft and is scanned by the light transmitted from four opposing grating patterns. The resulting alternating light and dark effects of the transmitted light beam are registered by photoelectric cells, and after modulating into electric signals these are counted by a pulse counter. Optical resolvers enable accuracies of up to ±1 second of arc with a resolution of 0.5 second to be obtained.

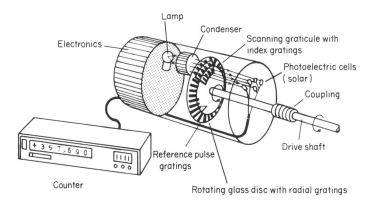

Fig. 2.30 Incremental angle encoder (optical resolver) (Heidenhain ROD 5)

2.2.2.1 Measurement of feed error (rotary/translatory movement) on a lathe

Figure 2.31 shows an inspection arrangement to determine the feed errors on a lathe using a laser interferometer and a resolver. Suppose, for example, that a thread is to be cut on the lathe; in such a case a fixed theoretical mathematical relationship will exist between the rotary motion of the machine spindle and the feed motion of the saddle. This is governed by the pitch of the thread to be produced. Hence, the objective of the test is to determine the degree to which this theoretical relationship is being observed. The resolver provides information with respect to the angular motion of the machine spindle. The feed motion of the saddle is measured with the use of a laser interferometer. The two motions are compared with each other and the deviation from the nominal relationship is obtained in terms of angular motion errors which are recorded on an analogue recorder.

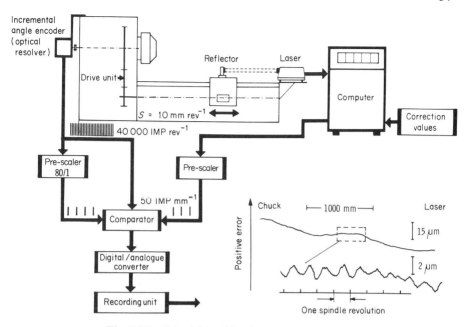

Fig. 2.31   Principles of feed-error measurement

The principle for producing an analogue error signal from the incremental impulses of the laser and resolver electronics may be seen in the diagram in Fig. 2.32. The voltage amplifiers convert the two incremental impulses (frequency f_1 from the resolver and f_2 from the laser interferometer) into the

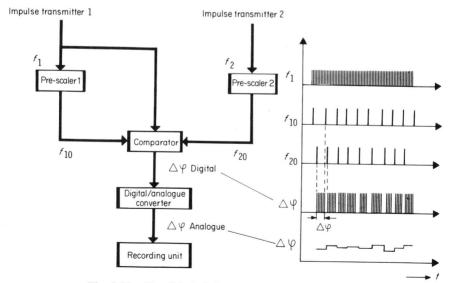

Fig. 2.32   Simplified digital error-transmission device

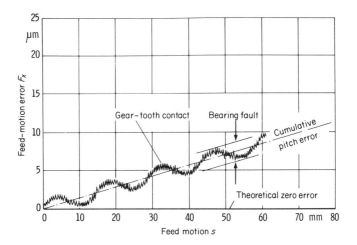

Fig. 2.33   Output trace of a feed-motion error

incremental impulses f_{10} and f_{20} in such a manner that $f_{10} = f_{20}$ when there is zero error. A counting gate in the comparater is opened by the impulses of frequency f_{20} and closed by the impulses of frequency f_{10}. The impulses from the resolver (f_1) registered during the time that the counting gate is open indicate in digital form the momentary relative phase position of the spindle with respect to the feed motions. This is converted by a digital/analogue converter into an analogue signal and recorded by a recording instrument over a time t.

A transmission error between the rotary and feed motion only occurs when the pulse counts, i.e. the phase relationships in the counting gate, fluctuate during one opening period.

Figure 2.33 shows the results of a feed motion error test. By examining the error pattern, conclusions may be reached as to the cause of the error.

2.2.2.2   *Measurement of rotation and feed motion errors on a gear-hobbing machine (rotary/rotary/translatory motion)*

Greater difficulties than those encountered on a lathe are experienced when, for example, the feed and rotary errors of a gear-hobbing machine are to be determined. Such a machine has a number of rotary and translatory axes whose movements are interrelated in accordance with fixed nominal mathematical relationships. In general, however, it is sufficient to examine three of these interdependent motions: workpiece rotation, hob rotation and feed motion of the hob slide. The inspection set-up is illustrated in Fig. 2.34. The rotary motions of the hob and workpiece are measured with optical resolvers (incremental angle encoders). The translatory hob feed is measured with a laser interferometer or a digital linear transducer.

A number of possibilities exist for interrelating the output signals and

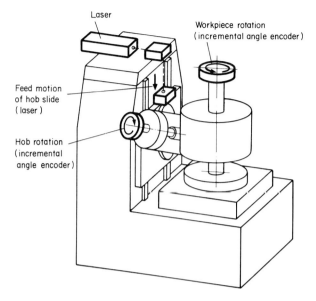

Fig. 2.34 Rotary- and feed-error measurement on a hobbing machine

determining the error; three of these are presented in the block diagram in Fig. 2.35:

(a) Measurement of the transmission error between the hob and table movements without considering the hob feed.

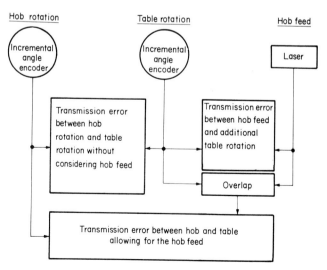

Fig. 2.35 Transmission errors in a gear hobbing machine

(b) Measurement of the transmission error between the hob feed and the additional table movement. (When hobbing straight spur gears, this error has no effect on the work accuracy.)

(c) Measurement of the transmission error between the hob and table rotations, taking the hob feed into account. When hobbing helical gear teeth, the error of the table rotation may compensate for the feed motion error of the hob slide under certain circumstances. For this reason the two error values are superimposed to obtain the actual hob rotation, and this is compared with the corresponding theoretical calculation. The difference between these values is then considered as the error.

Analogue measuring techniques are also applied to determine rotation errors with the use of a seismic rotational vibration detector. Figure 2.36 shows the construction of such a unit. The housing (base plate) is mounted on the face of the shaft under test. A rotor is suspended from a cross-spring joint, and constitutes the seismic mass. As a result of tuning to low frequency (natural frequency ≈ 1 Hz), the rotor mass rotates with uniform angular velocity because the non-uniform rotations of the shaft under observation at

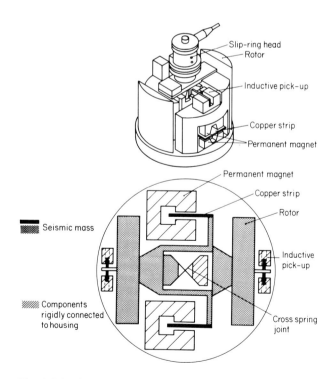

Fig. 2.36 Construction and principle of operation of a rotational vibration detector

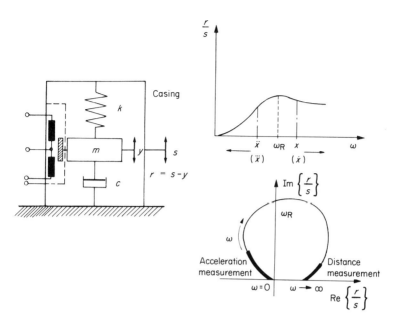

Fig. 2.37   Principles of a vibrometer (Budnick)

higher frequencies are not duplicated by the mass. Errors at frequencies which are lower than the tuned frequency are not detected by this method.

The rotary movement of the housing relative to the rotor is monitored by inductive displacement pick-ups and transmitted with the use of a slip ring head. The damping of the rotor, which must be proportional to its velocity—essential for its rapid stabilization—is obtained by eddy-current braking; to this end copper strips connected to the rotor are positioned between the poles of permanent magnets which is turn are fixed to the housing. The range of such devices is approximately between 1 and 400 Hz.

A method for observing uneven movements of feed drives in an analogue mode is by the application of a vibrometer (seismic linear oscillation detector). The construction and method of operation is shown in Fig. 2.37. A spring/mass system is tuned to frequency f_R. If the housing which is fixed to the machine component under investigation is subjected to movements having frequencies greater than $2f_R$ then the seismic mass will be quasi-stationary, so that the coil system is displaced relative to the plunge-type armature which is connected to the seismic mass. The test amplifier then indicates a signal proportional to the distance moved.

2.2.2.3   Measurement of two NC axes (translatory/translatory motion)
The axial movement of the saddle and the radial movement of the tool holder on the lathe shown in Fig. 2.38 are numerically controlled. When complicated shapes are to be produced—as, for example, tapered or spherical diam-

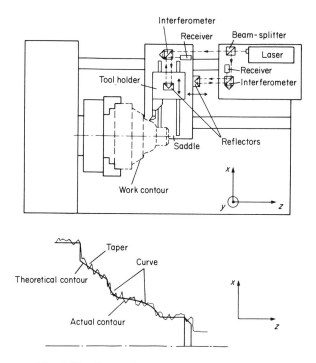

Fig. 2.38 Two-axis measurement on a NC lathe

eters—precise co-ordination of the feed motions is necessary. These requirements may be checked with the use of a laser, as shown in the diagram. The laser beam is divided into two components by a beam-splitter. While one part of the beam monitors the feed motion of the saddle in the axial direction, the other measures the radial feed movement of the tool holder.

From the measured feed motions a mini-computer calculates the actual contour of the work and compares this with the nominal work geometry which is stored in the computer's memory.

As the information with reference to the nominal movements of both axes is obtained from the interpolated outputs of the control unit, errors in the control unit itself are not detected by this method. One way to determine the overall machine accuracy, including that of the control unit, is by storing or generating from a punched paper-tape data input the nominal geometry of the component ($x_{nom} = f(z)$) in a plotter computer. These values can then be compared with the measured data ($x_{meas} = f(z)$) which will produce the machine motion errors, and this may also be done under load, i.e. while the component is being produced.

The errors may then be graphically presented by means of a digital plotter with a suitable magnification (see the lower part of Fig. 2.38).

3

THERMAL EFFECTS ON MACHINE TOOLS

When the thermal behaviour of metal-cutting machine tools is to be appraised, the only thermal deformations which are of interest are those which will lead to a relative displacement at the cutting point and thus have an influence upon the accuracy of the work being produced.

During an acceptance test, it is rare for the thermal behaviour of a machine to be separately appraised, e.g. by the introduction of defined heat sources;

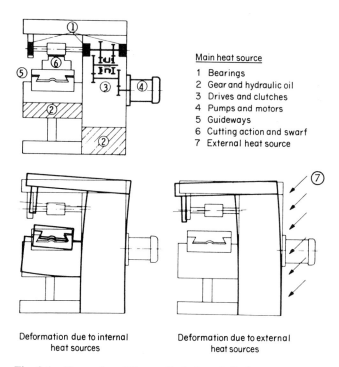

Main heat source

1 Bearings
2 Gear and hydraulic oil
3 Drives and clutches
4 Pumps and motors
5 Guideways
6 Cutting action and swarf
7 External heat source

Deformation due to internal heat sources

Deformation due to external heat sources

Fig. 3.1 Examples of thermally induced displacements on a milling machine

43

44

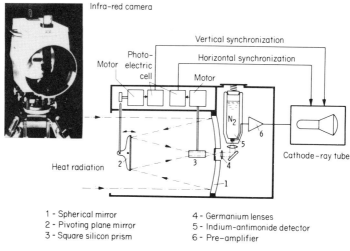

Infra-red camera

Vertical synchronization

Horizontal synchronization

Photo-
Motor electric
cell

Motor

N_2

Heat radiation

Cathode-ray tube

1 - Spherical mirror
2 - Pivoting plane mirror
3 - Square silicon prism

4 - Germanium lenses
5 - Indium-antimonide detector
6 - Pre-amplifier

Fig. 3.2   Infra-red   camera   for   determining   temperature
distribution

only the effect on the work accuracy is determined. Figure 3.1 shows various possible heat sources and their effect upon the displacement of the cutting point. As the thermal distortions of the shape of the machine components change, their effect upon the machining accuracy may be determined by measuring the geometric and kinematic behaviour (see Chapter 2) whereby the temperature distribution pattern over the whole machine is a parameter. The machine design can counteract the distortions at the cutting point by

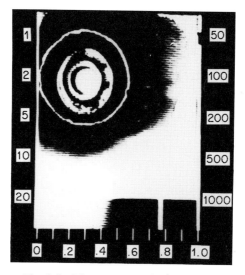

Fig. 3.3   Thermogram of a headstock

placing machine components which influence the work accuracy in 'thermally neutral areas'.

Infra-red photography which will enable isothermal patterns to be seen in addition to thermocouple elements and pyroelectric resistors (negative temperature-coefficient resistors) may be employed for the analysis of the temperature distribution. Figure 3.2 shows a suitable infra-red camera. The picture is scanned vertically and horizontally by a pivoting plane mirror (2), a rotating prism (3) and focused with a lens systems (4) on to a liquid-nitrogen-cooled and temperature-stabilized thermodetector. The electric signal is transmitted to a cathode-ray tube. The vertical and horizontal synchronization can be adjusted through the motors for the picture scan.

Figure 3.3 shows the thermogram of a headstock (viewed in the direction of the chuck). Areas of similar intensity of the black colouring indicate corresponding isothermal lines. The temperature is determined by means of a comparative scale (lower part of the picture) and calibration scales (left and right in the picture). By considering the surface finish, distance and colour of the object under examination, the absolute temperature may be found.

4

STATIC AND DYNAMIC BEHAVIOUR OF MACHINE TOOLS

The accuracy of work produced on metal-cutting machine tools is determined by the deviations at the cutting point from the required working movements between the tool and work. These deviations are caused—in addition to geometric and kinematic errors—by static and dynamic forces deforming all the components which are in the force-flux flow of the machine, such as machine frames, beds, slides, spindles, etc. The availability of modern methods of calculation has made it possible for the designer to obtain the required static stiffness. With regard to dynamic stiffness of systems, however, many interactions occur which can only be roughly estimated at the design stage. In particular, the lack of knowledge of the damping and stiffness conditions at interfaces, joints and couplings is today still a considerable factor of uncertainty when forecasting the dynamic machine characteristics.

For this reason, considerable reliance is placed upon metrological examinations when a precise determination is required of the stiffness or flexibility characteristics in an evaluation of a machine and when a scientific analysis with a view to improving measures is contemplated. Inadequate static stiffness of a metal-cutting machine tool is primarily responsible for errors in the form of the work produced (inadequate stability of size). In contrast, any uneven dynamic characteristics will lead to the generation of vibrations, the effects of which can lead to poor surface finishes on the work, increased machine and tool wear, as well as tool fractures and damage to both the workpiece and the machine. With this in mind, the flexibility characteristics of a machine in relation to changing load conditions must be regarded as a criterion of its performance capability.

4.1 Fundamentals of dynamic behaviour

Machine tools are assemblies of individual machine units, and therefore with respect to their dynamic behaviour they may be considered as multi-mass vibrators or oscillators. In many instances, the behaviour of a machine under

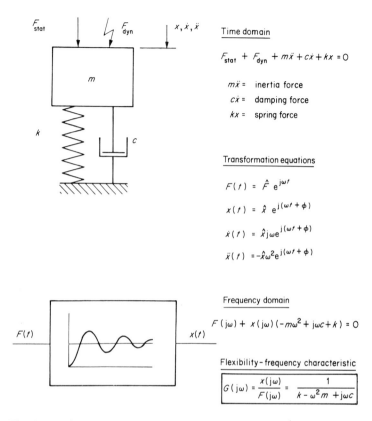

Time domain

$$F_{stat} + F_{dyn} + m\ddot{x} + c\dot{x} + kx = 0$$

$m\ddot{x} =$ inertia force
$c\dot{x} =$ damping force
$kx =$ spring force

Transformation equations

$$F(t) = \hat{F}\,e^{j\omega t}$$

$$x(t) = \hat{x}\,e^{j(\omega t + \phi)}$$

$$\dot{x}(t) = \hat{x}\,j\omega e^{j(\omega t + \phi)}$$

$$\ddot{x}(t) = -\hat{x}\omega^2 e^{j(\omega t + \phi)}$$

Frequency domain

$$F(j\omega) + x(j\omega)(-m\omega^2 + j\omega c + k) = 0$$

Flexibility – frequency characteristic

$$G(j\omega) = \frac{x(j\omega)}{F(j\omega)} = \frac{1}{k - \omega^2 m + j\omega c}$$

Fig. 4.1   Diagram of principles and transmission function of a
single-mass vibrator

the influence of dynamic loading may be approximated by considering the machine as a system of unconnected single-mass vibrators, enabling the dynamic machine characteristics to be demonstrated as a model consisting of examples of such single-mass vibrators. The law of motion of a single-mass vibrator under the influence of velocity-proportional damping is given in the upper part of Fig. 4.1. An overall view of the vibration effects can be obtained from the equation given in the lower part of the diagram for the frequency range. This presentation can be obtained directly from the differential equations, taking the transformation equations quoted into account, and will lead to the definition of the dynamic behaviour of machines in terms of the frequency-dependent flexibility (also known as receptance) $G(j\omega)$.

We are concerned here with the complex quotient due to the displacement x and the force F producing it. This frequency-dependent flexibility is fully defined for a single-mass vibrator by three variables in the system, i.e. static stiffness k, natural angular frequency ω_n and the damping ratio D (also designated $\zeta =$ zeta) (see Fig. 4.2) Apart from the equations which evaluate the

48

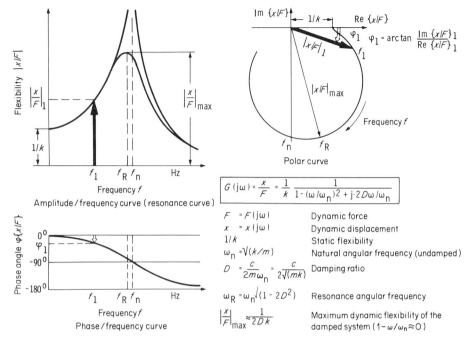

Fig. 4.2   Flexibility–frequency characteristic of a single-mass vibrator

variables ω_n and D, the diagram also presents the accepted method of graphical presentation of a frequency – response curve. The excitor frequency affects not only the degree of flexibility (amplitude – frequency curve) but also the time interval between the applied force and the consequential deformation (phase – frequency curve).

The polar curve shown in the right upper part indicates the amplitude – frequency and the phase – frequency characteristics. The distance between a point on the polar curve and the origin of the co-ordinates represents the degree of flexibility, whilst the angle between the polar vector and the positive real axis indicates the phase. Because of the passive characteristics of the system, there is always a time delay between the deformation and the applied force. This produces negative phase values in the phase – frequency characteristics and a need to define the frequency parameters of the polar curve in the clockwise direction, i.e. in the mathematically negative direction of rotation.

Theoretically, there are three typical angular frequencies which may be identified for a single-mass vibrator:[11,12]

$$\omega_n = \sqrt{\left(\frac{k}{m}\right)}$$

Natural angular frequency of an undamped system (90° phase lag between force and displacement).

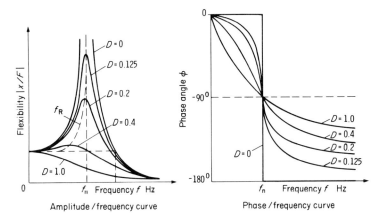

Fig. 4.3   Frequency characteristics of a single-mass vibrator

$\omega_{dn} = \omega_n \sqrt{(1 - D^2)}$ Natural angular frequency of a damped system. The amplitude of a freely oscillating system decays at this frequency.

$\omega_R = \omega_n \sqrt{(1 - 2D^2)}$ Resonance angular frequency of a damped system. At this frequency a real system experiences the maximum dynamic flexibility as a result of an harmonic excitation.

In general, the quantitive identification of these three frequencies is not necessary in machine-tool construction, because at the low system damping ratios which occur, $D \leqslant 0.1$, they coincide for all practical purposes.

Figure 4.3 provides a qualitative overview upon the effect of the damping ratio on the magnitude and phase of the flexibility of a single-mass vibrator. For an undamped system there is a resonance increase to infinity for a natural frequency f_n while the phase undergoes a sudden change from $0°$ to $-180°$. In real systems with damping ratios greater than zero, the resonance magnification remains in step with the flexibility, i.e. the resonance curve becomes broader as the damping ratio increases, whilst the phase change from $0°$ to $-180°$ occurs more gradually. For the dead-beat aperiodic limiting case when the damping ratio $D = 1$, there is no resonance increase on the flexibility.

4.1.1   Determination of system characteristics from measurements of the dynamic flexibility conditions

The 'static stiffness' k may be obtained directly from the measured flexibility – frequency characteristics as the reciprocal value of the flexibility at the frequency $f = 0$ Hz.

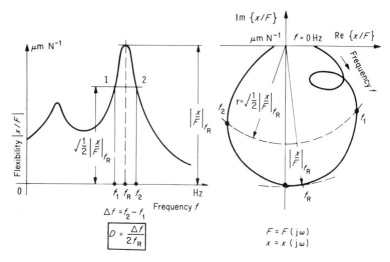

Fig. 4.4   Calculation of the damping ratio from the amplitude–frequency
curve and the flexibility polar curve by the $\sqrt{2}$ method

Machine tools consist of a multitude of mechanical components coupled together. Each of these spring/mass systems causes a resonance increase on the flexibility. In general, frequencies at which such resonance increases occur are known as resonance frequencies f_R. The position at which the absolute maximum flexibility takes place is therefore called the 'dominant resonance position' or point and the associated frequency is known as the 'dominant resonance frequency'.

The damping ratio D of a system, also known as the 'system damping', governs the resonance increase of the flexibility. Generally this can be determined by the so-called $\sqrt{2}$ method from the amplitude – frequency curve and the polar curve of the flexibility, as shown in Fig. 4.4. This technique can normally be used for every resonance increase of a measured flexibility – frequency relationship. For a boundary condition a judgement must be made as to whether the flexibility of the position being examined is, in addition, materially affected by neighbouring resonance points. The frequency difference necessary between neighbouring resonance points reduces as the damping ratio reduces, i.e. with a narrower spread of the resonance increase.

The determination of the damping ratio D for a system from the relationship between static flexibility and the maximum dynamic flexibility is only possible for pronounced conditions of single-mass vibrators, i.e. only when a single resonance increase of the flexibility occurs, as shown in Fig. 4.5.

A third method available for determining the damping ratio is by an evaluation of the decay curve of a system, as shown in Fig. 4.6. The system under observation is firstly harmonically excited into the resonance frequency which is of interest. After removing the excitation source, the decaying vibrations are plotted and the logarithmic amplitudes of the single periods are calculated.

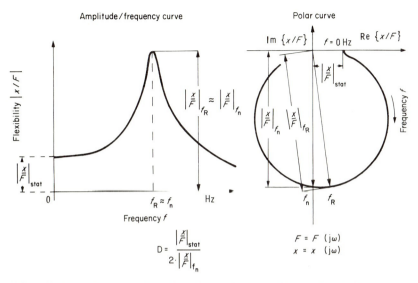

Fig. 4.5 Calculation of the damping ratio from the amplitude–frequency curve and polar curve of the flexibility

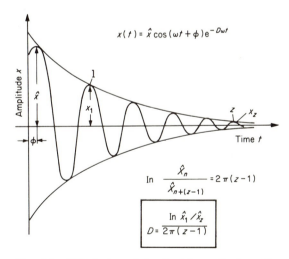

Fig. 4.6 Calculation of the damping ratio from the logarithmic decrement of the decay curve of a system

4.2 Types of vibrations and their causes

Even under continuous machining conditions relative movements take place between the work and cutting tool which will interfere with the required nominal motions, i.e. the feed and cutting movements. For the evaluation of these vibratory movements, it is usual to differentiate between externally

52

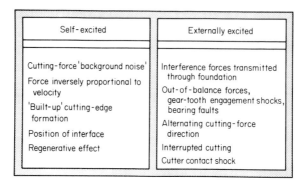

Self-excited	Externally excited
Cutting-force 'background noise' Force inversely proportional to velocity 'Built-up' cutting-edge formation Position of interface Regenerative effect	Interference forces transmitted through foundation Out-of-balance forces, gear-tooth engagement shocks, bearing faults Alternating cutting-force direction Interrupted cutting Cutter contact shock

Fig. 4.7   Vibrations in metal-cutting machine tools

excited or forced vibrations and self-excited oscillations. Figure 4.7 provides a summary of various vibrations under these classifications.

'Externally excited' vibrations occur as a result of, for example, interference forces transmitted through the foundations into the machine or as a result of damaged or imperfect machine elements in the form of unbalanced masses, bearing faults and gear-tooth engagement shocks. Varying cutting forces such as those experienced when milling due to the interrupted cut or the cutter impact shock also cause forced vibrations. The characteristic feature of such externally excited or forced vibrations is that the machine system will oscillate at the frequency of the excitation forces. This can result in particularly high amplitudes when the excitation frequency is near or even identical with a machine natural frequency.

Of course this is only valid for periodic excitations. In the case of impulse or transient excitations, the machine system will tend to oscillate at its natural frequency, whereby the vibration amplitudes will decay exponentially. In most cases forced vibrations can be reduced or eliminated by either removing the source of the interference or by periodically changing the exciter frequency in such a manner that it does not come close to any natural frequency of the machine system.

When milling, it is not possible to avoid the cutter impact shock—fundamental to this cutting process—and hence particular attention must be paid to the tendency of the machine towards vibrations induced by this type of excitation.

Under 'self-excited' vibrations, the machine system oscillates basically at one or more natural frequencies, when no external interference forces are acting.

One type of vibration which lies in the transition area between forced and self-excited vibrations is that of 'cutting-force background noise'. In this case, the externally excited contribution is due to the chip formation, i.e. the shear plane formation and the chip fracture. Self-excited vibrations occur as a result of the surface finish waviness induced by the externally excited vibrations, which lead to excitations ('self-excitations') after the work has made one revolution in the case of turning or when the next cutting tooth engages in the case of milling operations. The machine will tend to oscillate at its natural frequencies because it will naturally tend to produce its main dynamic flexure at these its resonance points.

The magnitude of this cutting-force spectrum which is generated by the cutting action itself is small, and consequently these vibrations are usually of no importance due to their small amplitudes. In grinding or polishing operations, however, they tend to produce minute relative movements between the work and tool perpendicular to the cutting surface, the effect of which may be detected by light reflections; vibrations caused by cutting-force 'noise' are in these cases an insoluble problem.

Another source of self-excited vibrations may be the reduction of the cutting force as a result of increasing cutting speed. In a similar manner to the stick-slip effect in slides, such an increasing cutting velocity in relation to reducing cutting forces indicates a negative damping effect and can lead to instability, i.e. to self-excited vibrations. However, such a vibration excitation is of little importance because this characteristic of reducing cutting forces/increasing cutting speed only applies to low cutting velocities which are not applicable to modern tool/work material combinations.

The formation of a so-called 'built-up edge' leads to changes in the cutting forces and consequently to corresponding displacements between the tool and work. This phenomenon, which occurs in the transition area between self-excited and forced vibration classifications, has so far not been sufficiently researched. It is doubtful whether the formation of built-up edges is directly related to a natural frequency of a machine. At the increased cutting speeds which are commonly used today, it is hardly possible to produce a built-up edge and hence this problem is no longer of major significance.

The vibrations which occur due to the position of interfaces are confined to vibration systems of higher orders with two closely adjacent natural frequencies and differing natural frequencies for such resonance points.

Self-excited vibrations caused by the regenerative effect are mainly a dynamic problem in metal-cutting machine tools. This excitation mechanism acts in direct relation to the 'background noise' of the cutting forces. Even when such cutting-force decays are very low, they nevertheless produce a waviness on the work surface limited by the finite stiffness of the machine. This waviness is particularly noticeable in the amplitude of motion at the resonance frequency of the machine. Noise-type force excitations also produce particularly large displacements at the resonance frequencies of the machine due to its greater flexibility at these frequencies. If cutting continues on this wavy surface, as is the case, for example, after every work revolution when turning, a dynamic excitation of the machine at its resonance frequencies is produced. The danger of a complete instability of the machining process is governed by the cutting conditions and the flexibility of the machine.

Figure 4.8 indicates the qualitative development trend of the vibration amplitudes in a face milling process in relation to the chip width for externally and self-excited vibrations. For a first approximation it may be assumed that

54

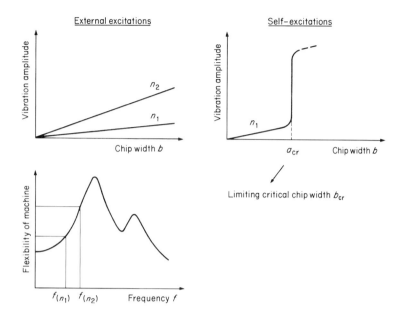

Fig. 4.8   Vibration amplitudes in relation to chip depth for external and
self-excitations

the dynamic forces at a cutting edge are proportional to its length. Therefore, the vibration amplitudes caused by the cutter impact shock are similarly proportional to the chip width, as may be seen on the left of the diagram. The constant of proportionality is mainly dependent upon the flexibility of the machine at the excitation frequency. The frequency of the displacement movement is identical with the excitation frequency or possibly its harmonic frequency for such externally excited vibrations.

For self-excited vibrations a similar linear rise in displacement values is observed for chip widths up to a certain critical limiting value. Beyond this critical chip width b_{cr} or critical depth of cut a_{cr}, an instability in the machining process occurs which is represented by the steep rise in the vibration amplitude curve. The frequency of these vibrations is always near the natural frequency of the machine.

In practice, it is often difficult to separate self-excited from externally excited vibrations. To this end, the decision logic presented in the block diagram of Fig. 4.9 can be of assistance. As soon as vibrations are detected, the machine is switched off.

If the vibrations, which are observed by sound, touch or electronic evaluation devices, do not stop, then we are faced in all cases with externally excited vibrations which are introduced from other equipment, e.g. through the foundations. If on the other hand the vibrations are eliminated when the machine stops, then we may conclude with some certainty that we are faced with other externally excited vibrations and their cause is to be found among the moving

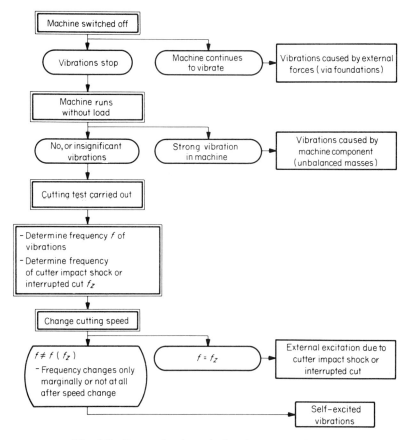

Fig. 4.9   Determination of vibration causes (Kegg)

components of the machine. By knowing the excitation frequency, the source of the excitation can usually be quickly determined.

If the frequencies of the vibrations change in proportion to changes of the input speed, then we are concerned with externally excited vibrations which are either initiated by the cutting process itself (e.g. cutter impact shock when milling, interrupted cuts, etc.) or have their origin in the drive elements. A calculation or measurement of the relevant excitation frequencies usually enables a definite judgement of the vibrations and their causes to be made.

Should the vibration frequency remain almost constant, even if the input speed is varied, then self-energized vibrations are acting, i.e. the vibration frequency corresponds to the natural frequency of the system.

4.3   Systematic description of the regenerative effect

As indicated in section 4.2, even a stable cutting process is not completely free from vibration effects due to the 'background noise' of the cutting forces.

These very small variations in cutting force still produce a waviness on the work surface, contingent upon the ultimate stiffness of the machine. When the tool cuts into this waviness, as in the case, for example, after one revolution of the work when turning, a dynamic excitation of the machine will ensue. The factors which determine the magnitude of the dynamic cutting-force variations are of the utmost importance for establishing the stability of the cutting process, i.e. its continuing progress. The dynamic cutting-force variations are among other factors dependent upon the changes in the cross-section of the chip at any given time, i.e. upon the variations in active cutting-tool edge length and chip thickness. The latter variable is governed on the one hand by the phase relationship between the previously produced waviness and the momentary dynamic displacement at the cutting edge due to the advancing depth of cut, and on the other hand the magnitude of the vibration amplitudes is influenced by the damping effect of the machine/cutting-process system.

In the case of a non-critical phase relationship or an adequate damping effect of the system, the cutting process will remain stable. If, however, the phase relationship is unfavourable, then when a particular active cutting-tool edge length is exceeded, defined as the critical chip width b_{cr}, a critical amplification of the machine/cutting-process system is reached at which the damping effect is inadequate; the vibrations will magnify and the cutting process will become unstable.

This procedure can be considered in a simplified form by looking at a turning process, as symbolically presented in the upper part of Fig. 4.10. Let us assume that a change in the cutting force occurs at time t_1 which produces a decaying natural vibration movement between the work and tool so that a harmonic surface contour is produced on the work (t_2 in Fig. 4.10). After one revolution of the work (t_3) the harmonic contour is cut by the tool. The resulting cutting-force variations excite the machine into new vibrations. Beyond a definite chip width, the damping effect of the system is inadequate for steadying the process, i.e. the cutting process becomes unstable (t_4). Because the new cut into the previously produced wavy surface serves to maintain the vibration pattern, this type of vibration effect is known as 'regenerative chatter'.

As mentioned earlier, only those relative movements which result in a change of chip thickness are of major importance, i.e. where the direction of movement is perpendicular to the machined surface of the work. In the lower part of Fig. 4.10, these relationships are presented in a simplified block diagram. The machining process is shown as a closed loop wherein the flexibility of the machine is represented in the forward branch and the cutting process in the return link. The behaviour of the machine is indicated by the so-called 'directional flexibility – frequency characteristic G_d (jω)' which takes the geometric conditions of the machining process under consideration into account, i.e. the relative position of the work and tool, the cutter geometry and the number of cutting edges of the tool being used. It represents the

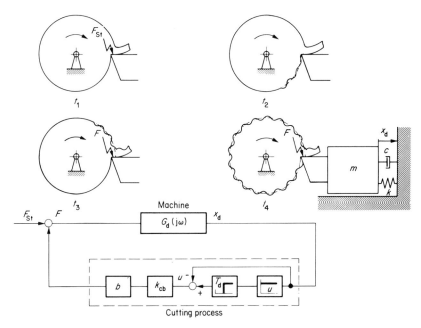

Fig. 4.10 Diagrammatic presentation of the regenerative chatter process

relative movements in the direction of the chip thickness between the work and tool caused by the dynamic cutting force.

The cutting process consists of the following components:

(a) The speed-dependent 'dead-time' period T_d. This is characteristic of regenerative chatter and covers the time between the production of a surface waviness and the renewed cutting of that wavy surface.
(b) The 'overlap factor' μ. This indicates the degree of overlap between one cut and the next (e.g. plunge cutting $\mu = 100\%$, thread cutting $\mu = 0\%$).
(c) The specific dynamic cutting stiffness k_{cb}. This is a material constant which is dependent upon the change in the cutting force as a result of dynamic changes in chip thickness.
(d) The chip width b, which is a measure of the active cutting-tool edge when considered in conjunction with the approach angle κ_M. From this we obtain, among other things, the absolute magnitude of the change in the dynamic cutting force. The limiting critical chip width b_{cr} is often used as a representative definition of the chatter-free cutting performance attainable for a particular machining process.

The effect of the machine and the cutting process upon the chatter conditions, i.e. the limiting chip width, may be obtained from a consideration of the stable state of the closed-loop machine-cutting process.[13] By applying the

58

'Nyquist criterion' for the stability loop, the following conditions must be satisfied for the frequency characteristic of the open loop:

$$\mathrm{Re}\{G_0(j\omega)\} \begin{cases} < 1 \text{ stable} \\ = 1 \text{ stability boundary} \\ > 1 \text{ unstable} \end{cases} \quad (4.1a)$$

$$\mathrm{Im}\{G_0(j\omega)\} = 0 \quad (4.1b)$$

As in most cases the overlap is complete, i.e. $\mu = 1$, we have:

$$G_0(j\omega) = G_d(j\omega)(e^{-j\omega T_d} - 1)k_{cb}b \quad (4.2)$$

where

$$e^{-j\omega T_d} = \cos(\omega T_d) - j \sin(\omega T_d)$$

Hence

$$\mathrm{Re}\{G_0(j\omega)\} = (\mathrm{Re}\{G_d(j\omega)\}[\cos(\omega T_d) - 1] \\ + \mathrm{Im}\{G_d(j\omega)\}\sin(\omega T_d))k_{cb}b \quad (4.3)$$

$$\mathrm{Im}\{G_0(j\omega)\} = (\mathrm{Im}\{G_d(j\omega)\}[\cos(\omega T_d) - 1] \\ - \mathrm{Re}\{G_d(j\omega)\}\sin(\omega T_d))k_{cb}b \quad (4.4)$$

By letting equation (4.4) equal zero, in accordance with the second part of the Nyquist condition (4.1b) we obtain:

$$\frac{\sin(\omega T_d)}{\cos(\omega T_d)} = \frac{\mathrm{Im}\{G_d(j\omega)\}}{\mathrm{Re}\{G_d(j\omega)\}} \quad (4.5)$$

Using

$$\tan\left(\frac{\alpha}{2}\right) = \frac{1 - \cos \alpha}{\sin \alpha}$$

we can write

$$\tan\left(\frac{\omega T_d}{2}\right) = -\frac{\mathrm{Re}\{G_d(j\omega)\}}{\mathrm{Im}\{G_d(j\omega)\}} \quad (4.6)$$

and so we obtain the relationship for the speed-dependent dead-time T_d, as $\omega = 2\pi f$:

$$T_d = \frac{1}{\pi f} \arctan\left(-\frac{\mathrm{Re}\{G_d(j\omega)\}}{\mathrm{Im}\{G_d(j\omega)\}}\right) \quad (4.7)$$

Because of

$$\mathrm{tg}(\phi) = \mathrm{arc\ tg}(\phi + m\pi), \quad \text{where} \quad m = 1, 2, 3, \ldots, \infty$$

equation (4.7) becomes in more general notation

$$T_d = \frac{1}{\pi f}\left\{\arctan\left(-\frac{\mathrm{Re}\{G_d(j\omega)\}}{\mathrm{Im}\{G_d(j\omega)\}}\right) + m\pi\right\} \quad (4.8)$$

In this condition the machine will tend to chatter as soon as a given limiting chip width b_{cr} is exceeded. If the speed of rotation and the number of cutting edges are considered, we obtain the following equation:

$$h = \frac{60}{zT_d} = \frac{60f[\text{He}]}{z\left(m - \dfrac{1}{\pi} \text{arc tan} \dfrac{\text{Re}\{G_d(j2\pi f)\}}{\text{Im}\{G_d(j2\pi f)\}}\right)} \quad \text{rev min}^{-1} \qquad (4.9)$$

where $m = 1, 2, 3, \ldots, \infty$

Using equation (4.9), for every possible chatter frequency series of critical machine speeds can be identified in relationship to the phase angle, described by the real and imaginary component of the directional flexibility – frequency characteristic $G_d(j\omega)$. The actual chip width b determines whether the machine will chatter at the various speeds. When b exceeds a limiting critical chip width b_{cr}, the cutting process becomes unstable. This may be derived from equation (4.1a) having regard to equation 4.3:

$$b_{cr} = \frac{1}{\text{Re}\{G_d(j\omega)\}k_{cb}\left\{\cos(\omega T_d) - 1 + \dfrac{(\sin(\omega T_d))^2}{\cos(\omega T_d) - 1}\right\}} \qquad (4.10)$$

Setting $\omega = 2\pi f$ we have

$$b_{cr} = \frac{1}{-2k_{cb}\,\text{Re}\{G_d(j2\pi f)\}} \qquad (4.11)$$

In order that the critical chip width b_{cr} is given a positive value, the condition for an unstable machining process is that the real part of the directional frequency characteristic should be smaller than zero ($Re < 0$). The absolute minimum for the critical chip-width value is then at the point of the maximum real component of the directional polar curve for the machine:

$$b_{cr\,min} = \frac{1}{2k_{cb}\,|\,\text{Re}\{G_d(j\omega)\}_{neg\,max}\,|}$$

The lowest usable critical chip width is therefore inversely proportional to the specific dynamic cutting stiffness and the maximum negative real component of the directional flexibility – frequency characteristic of the machine. Consequently, the smaller the negative real component of the directional flexibility – frequency characteristic, the less will be the tendency for the machine to be subject to self-excited vibrations, i.e. regenerative chatter.

The application of equations (4.9) and (4.12) permit stability charts with the appropriate chatter-frequency pattern to be prepared. An example of this is shown for a special machining operation in Fig. 4.11. Ranges of higher stability are indicated at higher rotational speeds ($a_{cr} > a_{cr\,min}$). This is due to the phase lag ε between the production of a wavy surface after one revolution

60

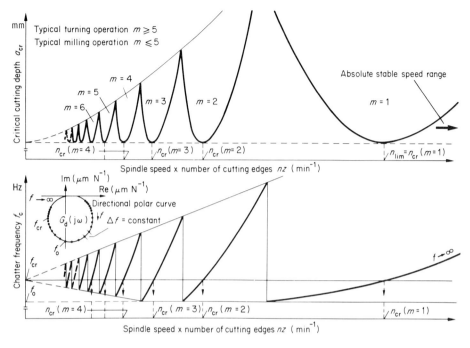

Fig. 4.11   Stability chart and chatter frequency pattern

and the wavy surface produced by subsequent cuts (which is dependent upon rotational speed and chatter frequency), causing significantly marked variations in chip thickness, as shown in Fig. 4.12. On the left of the diagram the vibration frequency is exactly a whole-number multiple of the rotational speed, causing the instantaneous vibrational movement of the tool to follow

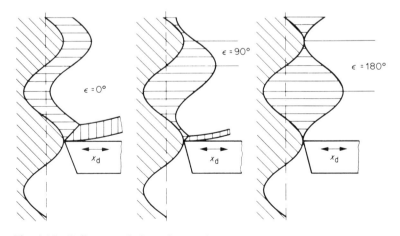

Fig. 4.12   Influence of phase lag angle ε upon variation in depth of cut

precisely the particular surface waviness; i.e. the phase angle ε between both vibrations is $0°$. In this case, there is no variation in chip thickness. The central diagram shows the two signal patterns with a phase lag of $\varepsilon = 90°$. Some variation in chip thickness is clearly noticeable. If the phase lag angle $\varepsilon = 180°$, then the waviness of the upper and under sides of the chip are exactly opposite; therefore under this condition maximum chip-thickness variation occurs.

As may be seen from the upper part of Fig. 4.11, the pitch between the minimum values on the curve increases as the speed increases. This is affected by the number of surface waves that it is possible to cut into by successive cutting edges with frequency f.

For typical turning operations the range of revolutions in the stability chart is $m \geq 5$ (where the curve peaks are closely spaced in respect to each other), but when considering typical milling operations, due to the relatively close spacing between the cutting edges, the usual value of $m \leq 5$ occurs. For this reason, it is hardly practical to take profitable advantage of the shape of the curves for turning work, but for milling operations this is a distinct possibility.

When cutting easily machinable materials with high cutting speeds, it is possible to work above the 'chatter curve' of the order $m = 1$. In this absolutely stable range of revolutions neither an externally excited nor a regenerative instability is experienced.

4.4 Factors influencing chatter behaviour

An objective evaluation of chatter on metal-cutting machine tools primarily requires the ability to reproduce the machine quality features which are to be measured. This in turn demands that all the boundary conditions which will affect a value judgement must also be reproducible. Experience has shown, however, that many factors which affect the chatter conditions of a machine follow theoretical principles, which in many cases are only known as general tendencies and are very difficult to establish by measurement. The influencing factors may be classified into machine work, tool and cutting-process groups. Figure 4.13 provides a summary of these individual parameters.

Of the many factors which have a material effect only a small number fall into the category for which the machine is responsible. All the others can be related to the workpiece, the tool or the cutting process itself.

When selecting and evaluating metal-cutting machine tools the effects upon the tendency to chatter of the individual influencing factors must be considered. These important conditions have been discussed with respect to their qualitative effect upon chatter conditions.[13]

4.5 Measurement of elastic deformations

When measuring the elastic deformations of machine tools it is usual to differentiate between static and dynamic flexibility conditions. However, static flexibility may also be considered as a special case of dynamic flexibility

Limiting critical chip thickness $b_{cr} = f$ (machine, workpiece / tool , cutting process)			
Machine		**Workpiece / tool**	**Cutting process**
Operating conditions	Directional orientation		
1. Foundations, installation conditions	Geometric influence in a given machining operation:	1. Work flexibility	1. Work material
		2. Work mass	2. Cutting geometry
2. Positioning of machine components	1. Direction of the dynamic cutting force due to setting angle and rake angle	3. Work clamping	3. Tool wear condition
3. Spindle speed		4. Work or tool diameter	4. Point radius
4. Slide and table movements		5. Tool flexibility	5. Cutting velocity
	2. Workpiece/tool configuration	6. Tool mass	6. Feed rate
5. Slack, backlash, non-linearity, pre-stressing, clamping conditions		7. Tool clamping	7. Hysteresis when determining limiting chip width
6. Operating temperature			8. Tool/work material combination
			9. Uneven distribution on multi cutting-edge tools
			10. Cooling and lubrication medium

Factors of influence

Fig. 4.13   Parameters which influence stability

with a frequency $f = 0$ Hz. This enables the magnitude of the static flexibility to be automatically included when the dynamic flexibility is determined. In recent years special excitation and evaluation analysis techniques have been developed to determine flexibility conditions. Current knowledge of examination and evaluation techniques permits accurate and economic tests of machines to be carried out.

Below, the simple testing techniques for the static flexibility conditions are presented in the first instance. This is followed by a description of types of power signal tests, analysis techniques, as well as motion, velocity, acceleration and force magnitude detectors, which are used for examining the dynamic conditions.

4.5.1   Measurement of static flexibility conditions

When considering static behaviour the criterion applied is the relative displacement at the cutting point resulting from a static load. Figure 4.14 shows a static loading test on a bed-plate milling machine. The force is passed through a force-measuring transducer and the relative displacement is determined by a dial indicator.

To determine the overall static flexibility conditions, the static flexibility of the milling spindle in relation to the machine table is determined for all three co-ordinate directions in turn. The displacement at the cutting point is the

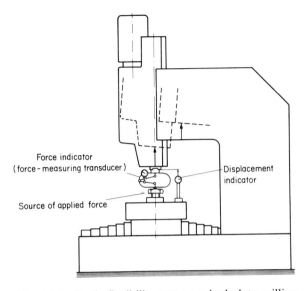

Fig. 4.14   Static flexibility test on a bed-plate milling
machine

sum of the deformations of all components and at the interfaces which are in
the force-flux circuit.

Figure 4.15 gives a deformation analysis of a boring and milling machine.
This shows the extent to which the individual machine components contribute
to the displacement at the cutting point. The magnitudes are determined by
analysing the deformations of the individual machine components—based
upon the overall machine deformation—and projecting these into their effect

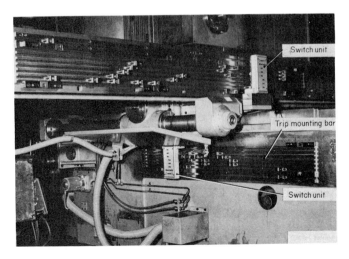

Fig. 4.15   Trip carrier for table movement control (Heller)

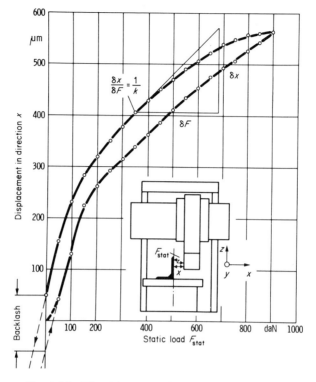

Fig. 4.16 Elasticity curve for a vertical boring mill

at the cutting point. A complete description of the static behaviour for a given direction of loading is only possible if an elasticity curve is produced.

Figure 4.16 shows the elasticity curve of a single-column vertical boring mill when loaded in the direction of feed. After the backlash has been taken up, the flexibility of the system progressively reduces. A hysteresis is formed when unloading takes place due to the changed contact conditions at the interfaces.

There are no standard procedures for the establishment of the static behaviour, but the static flexibility is a contributory factor when the work accuracy is evaluated (see section 6.1.2).

4.5.2 Measurement of dynamic flexibility conditions

In terms of control technology, a machine tool is regarded as a transmission or transfer element. The transfer behaviour of a machine may therefore be defined in the form of frequency response curves, which are known as flexibility – frequency response curves in this case.

Flexibility – frequency response curves provide the basis for an evaluation of the machine under test with respect to the stability of the cutting process

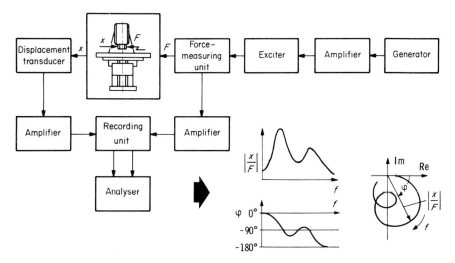

Fig. 4.17   Basic set-up for dynamic flexibility tests on machine tools

and to the likelihood of the occurrence of externally excited vibrations. In order to produce the flexibility – frequency response curves by a testing procedure it is necessary to excite the machine by forces with alternating direction and to measure the displacements which occur as a reaction to these forced excitations.

Figure 4.17 shows the basic set-up necessary for such tests. A generator produces an electric signal which is amplified and fed into an excitation unit. The alternating force exciter acts on the machine through a force-measuring unit. The signals from the force-measuring unit and from a displacement transducer which indicates the resulting movement are fed into the recording unit. The subsequent analysis of the signals and further processing into a flexibility–frequency response characteristic in the form of amplitude and phase curves and/or polar curves is made with the aid of a Fourier analyser. The results can usually be obtained graphically from an x/y plotter.

When analysing the dynamic behaviour of a machine, sinusoidal, stochastic and pulse or aperiodic input signals are employed. Because of these various types of signals a variety of excitation and analysis techniques that may be applied are available. Figure 4.18 presents these various types of signal and provides a summary of the classifications performance capabilities and applications of various types of excitation. Each of the methods has some shortcomings under certain circumstances, but these can be overcome by applying one of the other two techniques.

A particular difficulty is encountered when the flexibility–frequency characteristics have to be established where the machine units are in their feed motion or the work spindle is rotating. According to Fig. 4.18 only stochastic and aperiodic test signals can be used under these circumstances,

Test power signal forms	Exciter	f_{max} (Hz)	max F_{dyn} (N)	max F_{stat} (N)	Costs		Machine condition	Analysis of weaknesses
					Capital	Running		
Sinusoidal	Electrodynamic relative absolute exciter	3000	20	70	Low	High	Stationary	By interpreting sinusoidal signals
	Electrohydraulic relative exciter	800	1500	7000				
Stochastic	Electromagnetic relative exciter	1000	500	200	High (Fourier analyser)	Low	Rotating units	
	Electrohydraulic absolute exciter	300	2000	–			Linear moving units	
Aperiodic	Impulse hammer	2500	5 Hz^{-1}	–			Rotating and linear moving units	By modal analysis of stochastic and aperiodic signals
	Intermittent impact generator	(2500)	(5 Hz^{-1})	4000			Stationary	

Fig. 4.18 Features and applications of various test signals and methods of excitation

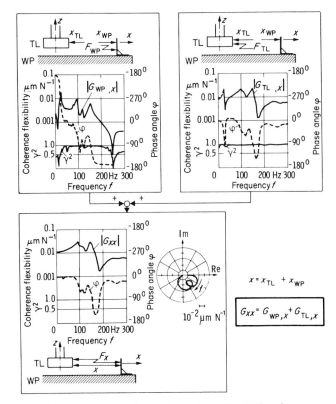

Fig. 4.19 Determination of relative flexibility–frequency response curves with rotating work spindle and under-feed motion

because the testing time under sinusoidal excitations is too long, making it impossible to imitate a stationary condition of the machine.

A reasonably economic testing procedure for establishing flexibility – frequency response curves for rotating and linear moving machine components is to excite the machine with a force pulse which is obtained from a hammer-type absolute excitation unit. Because we are then dealing with an absolute excitation, and given that the displacement is also measured with an absolute displacement transducer, it is necessary to measure four frequency characteristics which will provide the relative flexibility – frequency response characteristic G_{rel} by addition in the proper relationship.[14]

In an actual practical test, however, the procedure is that the relative displacement between the tool and work is measured as shown in Fig. 4.19. If the force excitation is applied on the workpiece, then absolute linear transducers are fixed to it and the tool in opposite directions so that the feed motion can take place during relative displacement measurement (Fig. 4.19, upper left). As a result of the particular type of excitation, the test periods are so short that higher feed velocities can also be examined.

Because the contact time between the exciter (hammer) and the component is short when this type of signal is used, a direct force application on rotating machine units is also possible. The relative displacement between the now-stationary table and the rotating work spindle is measured with a relative linear transducer. Contacting-inductive linear transducers which slide on a line of measurement have been found to be particularly trouble-free.

4.5.3 Types of exciters

The fundamental classification for exciters is between absolute and relative types, as shown in Fig. 4.20. Absolute excitation is the term used when the

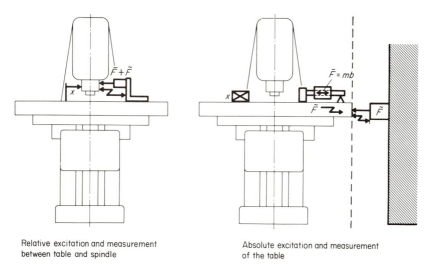

Relative excitation and measurement
between table and spindle

Absolute excitation and measurement
of the table

Fig. 4.20 Comparison between excitation and measurement techniques

68

force is applied to only one machine component. This may be achieved with a seismic mass or by supporting a relative exciter on a reaction point outside the machine.

For a relative excitation the force is applied between the tool and the workpiece.

4.5.3.1 *Electrodynamic exciters*
For the examination of smaller machines and models, electrodynamic exciters as shown in Fig. 4.21 are found to be useful. Only comparatively low forces are available (<30 N), but excitation frequencies up to 3000 Hz may be reached.

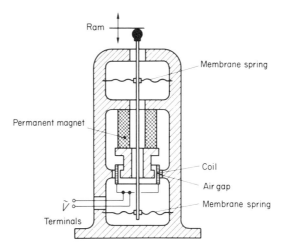

Fig. 4.21   Diagram of an electrodynamic relative
exciter

4.5.3.2 *Electrohydraulic relative exciters*
Electrohydraulically operated exciters have been found to be extremely effective for almost all sizes of machine. Their particular advantages are derived from their small and easily handled overall size.[15]

Figure 4.22 shows a sectional drawing of an electrohydraulic relative exciter. The oil flow from a hydraulic control unit is fed through a servo-valve to each of the piston faces alternately in accordance with the associated electrical signals. The static pre-load is applied over the rear piston surface; within a limited range, this may be varied by a DC voltage which is superimposed on to the electrical alternating current signal. The effective force is measured with a wire strain gauge.

Dynamic forces with frequencies up to 800 Hz and maximum exciter forces in accordance with the curve shown in Fig. 4.23 may be obtained, depending upon the stiffness of the system under examination. The static pre-load may be up to 7000 N.

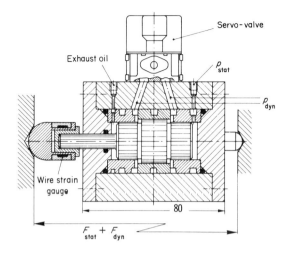

Static pre-load: $p_{stat} \leqslant 7000 \, N$
Alternating force: $p_{dyn} \leqslant 1500 \, N$
Frequency range: $0 - 800 \, Hz$

Fig. 4.22   Electrohydraulic relative exciter

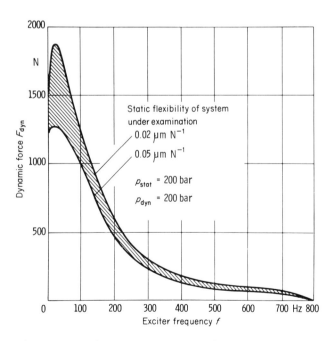

Fig. 4.23   Maximum power pattern of an electrohydraulic relative exciter

70

4.5.3.3 Electrohydraulic absolute exciters

For the testing of moving machine components and for such machine units where the support of the exciter is difficult, an absolute exciter has been developed which works in reverse to the seismic principle. Figure 4.24 shows the layout of a electrohydraulic absolute exciter. A freely moving mass is placed on a ram and is reciprocated by pressurized oil which is controlled by a servo-valve. The reaction to the force is transmitted to the object under test by the ram. The positional control system shown on the left of the diagram retains the mass in the centre.[16]

A disadvantage of this exciter is that as a result of the seismic principle no pre-load can be applied. For this reason additional pre-loading devices will be required.

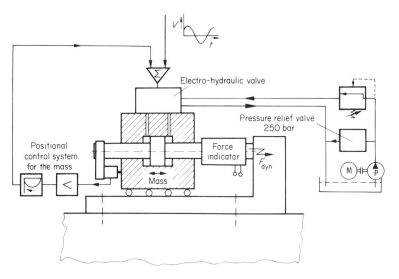

Fig. 4.24   Principles of the electrohydraulic absolute exciter

4.5.3.4 Electromagnetic relative exciters

Electromagnetically operated exciters are used to enable the influence of the spindle speed upon the dynamic flexibility behaviour to be determined.[17] Figure 4.25 illustrates the principle of such an exciter. The magnetic flux of a U-shaped electromagnet is looped through the rotating tool (for milling). To avoid eddy-current losses, the rotating section of the dummy work or tool is made out of iron laminations. A magnetic induction is produced by two separated coils which are charged with alternating and direct current respectively; this in turn produces static and dynamic forces. The exciter forces are monitored by piezo-electric load cells. The linear displacement is measured by eddy-current recorders which are not in physical contact. This type of exciter enables dynamic forces up to 130 N with maximum static forces of 2000 N to be generated within a frequency range up to 1000 Hz.

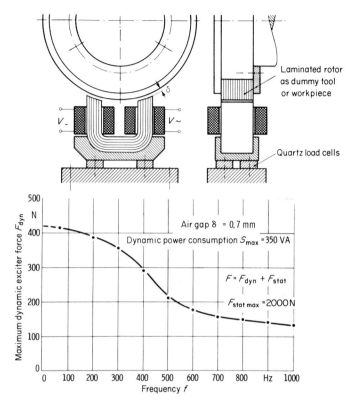

Fig. 4.25 Electromagnetic exciter; power curve based on frequency

4.5.3.5 Impulse hammers

A hammer-type absolute exciter, such as that shown in Fig. 4.26,[18] is used to produce an impulse force. A piezo-quartz load cell is used to measure the

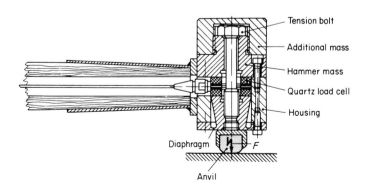

Fig. 4.26 Hammer for producing impulse signals for tests

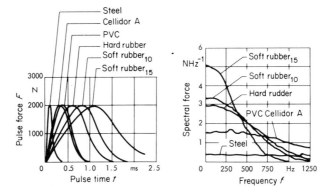

Fig. 4.27　Effect of anvil materials upon the pulse time and spectral force

force which is applied to the object under examination by means of an inter-changeable anvil. Alternatively, the force may be determined with strain gauges on a similarly designed hammer.

By employing anvils made from materials with varying hardness, the frequency range of the excitation spectrum may be varied to suit a given application, as shown in Fig. 4.27.

The eccentric-mass exciter shown in Fig. 4.28 is an absolute type which in the main only produces sinusoidal forces. The eccentric masses which rotate in opposite directions have the effect that the instantaneous x components of the force cancel each other whilst their corresponding y components are additive. The frequency of the excitation is governed by the rotational speed of the meshed gear wheels. A disadvantage of this type of exciter is that the force is

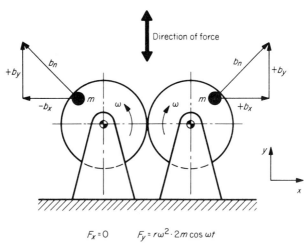

$$F_x = 0 \qquad F_y = r\omega^2 \cdot 2m \cos \omega t$$

4.28　Principle of an eccentric mass exciter

proportional to the square of the frequency and it is difficult to vary the magnitude of the force (variation of the mass m or its radial position).

4.5.4 Force measurement

In the main, the magnitude of the force is determined by strain gauges or by quartz load-cell transducers. With strain gauges, the expansion of the object being measured is determined which is proportional to the stress on its cross-sectional area within the range of validity of Hooke's law. Strain gauges operate by indicating the changes in the electrical resistance of suitable materials (constantan, semi-conductors) when these are stretched. The strain gauges are connected in the form of an electrical bridge; with the aid of measuring amplifiers the imbalance of the bridge resulting from the changes of resistance in the strain-gauge materials due to their expansion is recorded.

Figure 4.29 shows three possible bridge circuits and in diagrammatic form the mounting of the strain gauges in accordance with particular applications.[19] The output voltage of the bridge V_0 due to the expansion is given by:

$$V_0 = \frac{\varepsilon B K}{4} V_s$$

where V_0 = output voltage of bridge
V_s = supply voltage to bridge
ε = $\delta l / l$ = expansion of testpiece
B = bridge constant (see lower part of Fig. 4.29)
K = magnification factor
$\quad K$ = s for constantan strain gauges
$\quad K \approx 100$ to 200 for semi-conductor strain gauges
μ = transverse contraction ratio

In piezo-electric load cells the electrical charges in the crystal change under load. This change in the charge is measured with the aid of so-called charge amplifiers. Figure 4.30 shows a diagrammatic presentation of quartz crystal and the designs of commercially available quartz load cells.

4.5.5 Displacement, velocity and acceleration measurement

Figure 4.31 shows the fundamental principles of displacement, velocity and acceleration measurement devices. The respective output signals are as follows:

Proportional to displacement:
(a) Strain gauges, expansion of soft spring.
(b) Change in inductance of coils (plunger-type armature).
(c) Change in capacitance, without physical contact on object under test.

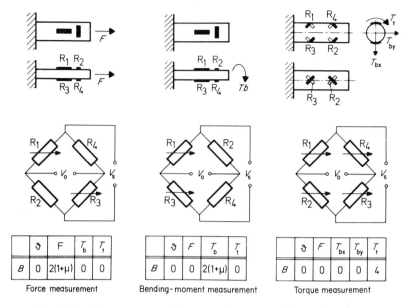

Fig. 4.29 Measurement of forces and torques with strain gauges

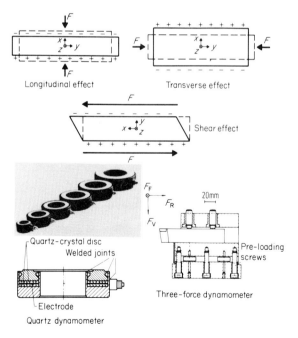

Fig. 4.30 Force measurement with piezo-electric quartz load cells

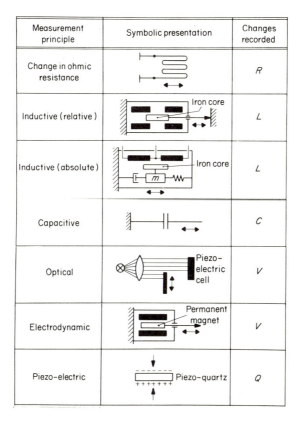

Measurement principle	Symbolic presentation	Changes recorded
Change in ohmic resistance		R
Inductive (relative)	Iron core	L
Inductive (absolute)	Iron core, m	L
Capacitive		C
Optical	Piezo-electric cell	V
Electrodynamic	Permanent magnet	V
Piezo-electric	Piezo-quartz	Q

Fig. 4.31 Fundamental principles of devices for measurement of displacement, velocity and acceleration

(d) Optical methods, lasers, photodiodes, holography.
Proportional to velocity:
  Electrodynamic principles.
Proportional to acceleration:
  Measurement of the reactive force of a moving mass. The force is measured with a piezo-quartz transducer or a strain gauge at the mounting position of the mass.

Figure 4.32 shows size comparisons of various displacement, velocity and acceleration transducers. Care must be taken to ensure that the mass of the transducer is small in relation to the mass of the system being examined so that the transducer itself has only a negligible influence on the system.

Figure 4.33 shows performance comparisons of various transducers. The selection criteria with regard to a given examination requirement are mainly the measurement range, sensitivity, permissible frequency range and the mass.

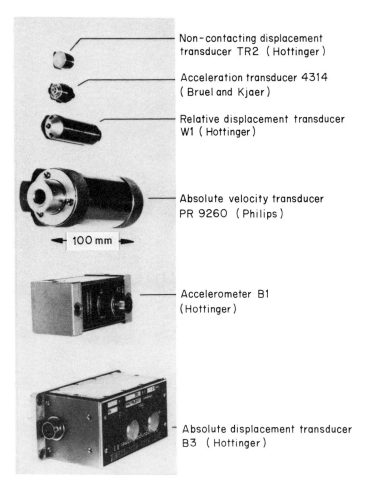

Non-contacting displacement
transducer TR2 (Hottinger)

Acceleration transducer 4314
(Bruel and Kjaer)

Relative displacement transducer
W1 (Hottinger)

Absolute velocity transducer
PR 9260 (Philips)

◀ 100 mm ▶

Accelerometer B1
(Hottinger)

Absolute displacement transducer
B3 (Hottinger)

Fig. 4.32   Various types of transducer

4.6   Analysis techniques

4.6.1   Analysis when using sinusoidal input signals

When analysing the transmission ratio of a machine using sinusoidal input signals, the frequency is changed steadily or in discrete steps within the required range. The displacement—or force—amplitude relationship and the phase lag between the input and output signals is then measured in the vibrating state. The results are presented in the form of Bode diagrams or polar curves.

The force and displacement signals may be evaluated automatically with the aid of a transmission function measuring unit, the principles of which are clarified in the block diagram in Fig. 4.34. The output signals are frequently

Transducer	Measurement principle	Measured variable	Relative/ absolute	Active/ passive	Maximum range	Contacting	Frequency range Hz	Mass kg
Hottinger W1T	inductive	X	relative	passive	±1 mm	yes	0–2000	0.080
Hottinger W1O	inductive	X	relative	passive	±10 mm	yes	0–2000	0.110
Hottinger Tr1	inductive	X	relative	passive	0.45 mm	no	0–1500	0.001
Hottinger B3	inductive	$\dot{X}$	absolute	passive	±2 mm	yes	10–300	1.200
Philips PR 9260	electrodynamic	$\dot{X}$	absolute	active	±3 mm	yes	12–500	0.500
Brüel & Kjaer accelerometer 4319	piezo-electric	$\ddot{X}$	absolute	active	$\dfrac{200\,g^{*}}{(2\pi f)^{2}}$	yes	3–10 000	0.030
Astatic pendulum	inductive	X	absolute	passive	±2 mm	yes	1–400	3.000
DISA reactance transducer	capacitance	X	relative	passive	a few mm	no	0–100 000	0.100

Fig. 4.33 Data on various transducers

$*f$ = Frequency (Hz).

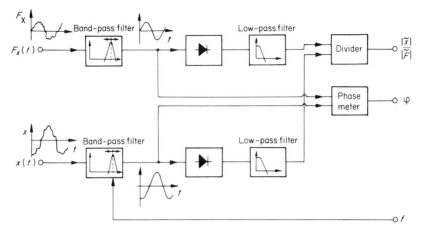

Fig. 4.34 Simplified block diagram of a transmission function measuring unit

disturbed by interference signals of varying strength or by higher harmonics. For this reason, the interfering signals are filtered out with the aid of self-adjusting band-pass filters (tracking filters) or by modulation with pure harmonic signals of the analysing frequency. The frequency response curve is obtained by dividing the rectified and smoothed displacement signals by the input force signals.

4.6.2 *Analysis when using stochastic and aperiodic input signals*

Due to the rapid progress in the development of digital computer technology with regard to capability, size, cost and speed of operation, digital analysis devices are mainly used today. They enable even signals which are irregular and of very short duration to be simply and speedily evaluated.

Figure 4.35 shows the hardware configuration of a digital Fourier analyser which has a 32k mini-computer at its core. The input and output signals, i.e. the force and displacement signals, which are taken directly from the machine are digitalized and are transformed into the frequency range by a fast-Fourier transformation (FFT) program. In less than one minute the complete frequency characteristic appears on a screen and by variations of the program all imaginable operations and output forms may be obtained, e.g. amplitude and phase curves against a linear or logarithmic frequency axis, polar diagrams, power and cross-power spectrum, coherence function, any necessary corrections for the transducer or tape frequency characteristics, storage of results and comparison with new readings, smoothing of functions, frequency distribution of signals, etc.

The results obtained may be transmitted directly to a 'hard-copy' paper plotter, to a punch for production of a punched paper tape or to a 'floppy disc' for storage so that the results of a very wide range of readings are available at exceptionally short notice.

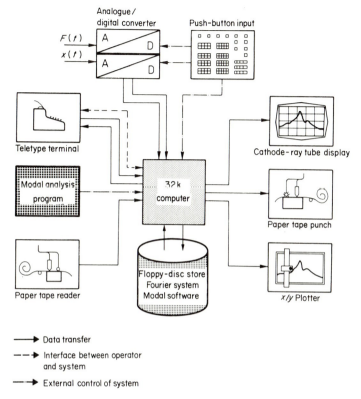

Fig. 4.35   Hardware configuration of a Fourier analyser (Hewlett-Packard)

The system presented in Fig. 4.35, which is shown in its actual form in the photograph in Fig. 4.36, enables a wide range of additional evaluation exercises to be performed, e.g. the Modal analysis described in Fig. 4.44. As this computer system is transportable the evaluation and the presentation of the results can be obtained at the actual place where the readings are taken.

Figure 4.37 shows a flow diagram for calculating the flexibility – frequency response curves with a Fourier analyser. From this, it may be seen that the time signals of the system's input (force) $x(t)$ and of its output (displacement) $y(t)$ are firstly digitalized, then transformed into the frequency range and then multiplied by a correction function. Such a correction function is necessary for stochastic test signals because of the errors which may possibly be introduced as a result of the signal reception time being finite.[20] The calculation of the auto-power or cross-power spectrum follows, which is necessary to determine the flexibility – frequency response curve and the coherence function. The coherence function is a measure of the uncertainties of measurement. It will lie between 0 and 1 depending upon the magnitude of the uncertainty. When the value is 1 the reliability of the measurement is at its optimum. Small values of the coherence function indicate that the interference signal level is

Paper tape punch   Modal analysis   Cathode-ray Push–button x/y Plotter
\   control   tube display input

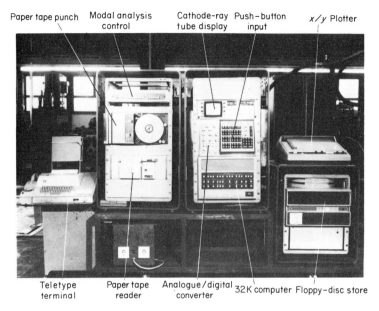

Teletype   Paper tape   Analogue/digital 32K computer Floppy–disc store
terminal   reader   converter

Fig. 4.36   Practical layout of the hardware configuration of a Fourier
analyser

very high in comparison with the useful signal level, and the system may be non-linear.

The mathematical fundamentals for calculating frequency characteristics into auto- and cross-power spectrums are outlined in Fig. 4.38.

4.6.3   Reduction of interference signal effects

When the frequency characteristics are measured for moving machine components, special problems arise as a result of any vibrations which are not initiated by the exciter. Examples of these are interference vibrations caused by the feed mechanism (non-uniformity, hunting) or by the rotating work spindle (bearing faults, roundness errors, imbalances).

All faults which are of a periodic nature may be completely nullified by the following technique. The interference signal (e.g. an error in the line of measurement on a rotating machine component without a force excitation) is recorded within the time base and deducted in the proper phase relationship from the signal recorded of the force excitation.

The upper diagram in Fig. 4.39 shows the false reading x_1 obtained from a pulse-shaped test signal on a time base due to an interference signal. Below this the error in the line of measurement x_2 is shown. The subtraction $x_1 - x_2$ in the proper phase relationship produces the corrected output signal x_3.

As may be seen from the upper right section of Fig. 4.39, the calculation of

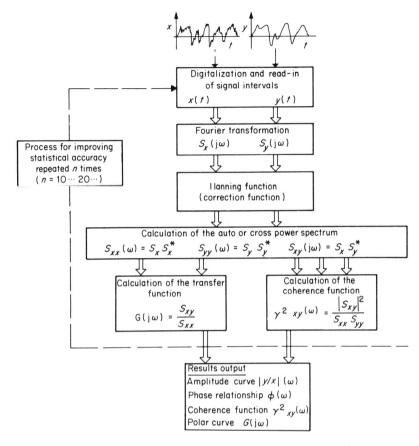

Fig. 4.37 Flow diagram of the program for calculating frequency characteristics from stochastic signals

a flexibility – frequency response curve based upon an uncorrected output signal x_1/F gives a result which has very little meaning because there is no correlation between the force and output signal in the frequency range below 200 Hz, as shown by the coherence function. A considerably improved evaluation is obtained by the application of the corrected output signal x_3 (lower right section of Fig. 4.39). Periodic interference vibrations (e.g. background noise of the measuring system) must not be given prime consideration in the case of impulse excitations because of their low spectral energy density when compared with sinusoidal or stochastic force excitations. A reduction of the influence upon the signal by the interference signal may be obtained through a variety of techniques, as, for example, the employment of several measurement probes and the utilization of suitable window functions when evaluating the output signals.

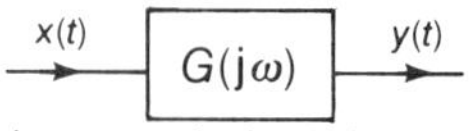

For a complex frequency characteristic $G(j\omega)$ of a general system we have:

$$G(j\omega) = \frac{\mathcal{F}\{\text{output signal}\}}{\mathcal{F}\{\text{input signal}\}} = \frac{\mathcal{F}\{y(t)\}}{\mathcal{F}\{x(t)\}} = \frac{\displaystyle\int_{-\infty}^{\infty} y(t)e^{-j\omega t}\,dt}{\displaystyle\int_{-\infty}^{\infty} x(t)e^{-j\omega t}\,dt} = \frac{E_y(j\omega)}{E_x(j\omega)}$$

where $\quad E_x(j\omega)$ = complex energy spectrum of the input signal
$\qquad\quad E_y(j\omega)$ = complex energy spectrum of the output signal

For the time function we have: $x(t) = t(t) \equiv 0 \begin{cases} t \leq t_1 \\ t \geq t_2 \end{cases}$; hence:

$$G(j\omega) = \frac{E_y(j\omega)}{E_x(j\omega)} = \frac{\displaystyle\int_{t_1}^{t_2} y(t)e^{-j\omega t}\,dt}{\displaystyle\int_{t_1}^{t_2} x(t)e^{-j\omega t}\,dt}$$

After dividing the complex energy spectrum by the integration time, we obtain the complex power spectra:

$$S_x(j\omega) = \frac{1}{t_2 - t_1} E_x(j\omega) = \text{complex power spectrum of the input signal}$$

$$S_y(j\omega) = \frac{1}{t_2 - t_1} E_y(j\omega) = \text{complex power spectrum of the output signal}$$

$$G(j\omega) = \frac{S_y(j\omega)}{S_x(j\omega)} = \frac{\text{Re}\{S_y(j\omega)\} + j\,\text{Im}\{S_y(j\omega)\}}{\text{Re}\{S_x(j\omega)\} + j\,\text{Im}\{S_x(j\omega)\}}$$

After expansion with the complex conjugate denominator we obtain the following for the frequency characteristic:

$$G(j\omega) = \frac{S_{xy}(j\omega)}{S_{xx}(\omega)} = \frac{\text{Re}\{S_{xy}(j\omega)\} + j\,\text{Im}\{S_{xy}(j\omega)\}}{S_{xx}(\omega)}$$

where $\quad S_{xx}^*(\omega) = S_x(j\omega)\,S_x(j\omega) = \text{auto-power spectrum (real)}$
$\qquad\quad S_{xy}^*(j\omega) = S_x(j\omega)\,S_y(j\omega) = \text{cross-power spectrum (complex)}$

Fig. 4.38   Mathematical fundamentals for calculation of frequency characteristics

A particularly effective technique to reduce the interference signals has been presented.[18] The upper left section of Fig. 4.40 shows a flexibility – frequency response curve which is falsified in the whole frequency range by an interference signal. If the inverse Fourier transformation is produced, then the system output from a Dirac delta is obtained, i.e. the weighted function. As the system under investigation may be considered to be passive, the assumption can be made that all the signals which may be related to the

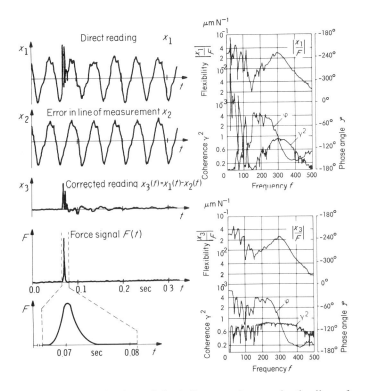

Fig. 4.39 Elimination of the influence of errors in the line of measurement

system will decay to an insignificant magnitude after a time T. This time T may be estimated, based upon the frequency content and the system damping. With a modified Hanning window function (Fig. 4.40, right centre) the signals which are constant up to the end of the time window and do not correspond to the passive characteristics of the system are suppressed. The final step is to produce the frequency characteristic by a reverse transformation of the evaluated system output, which will define the system characteristics with the interference signals almost eliminated.

As an example of the effectiveness of this procedure, Fig. 4.41 shows the flexibility – frequency response curve which is influenced by stronger interference signals due to a higher spindle speed than that used in Fig. 4.40. A meaningful evaluation of the output signal is only possible to a limited degree, as may clearly be seen from the polar curve in the right section of Fig. 4.41. By applying the technique to minimize the effect of the interference signal, the frequency characteristic shown in the lower part of the diagram is calculated.

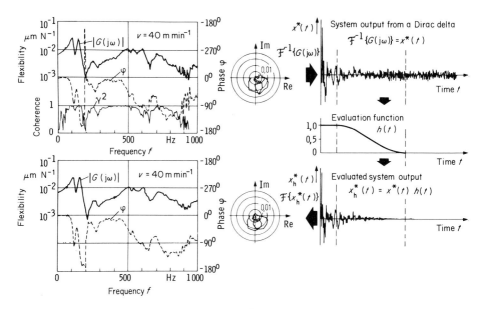

Fig. 4.40 Reduction of the effect of an interference signal

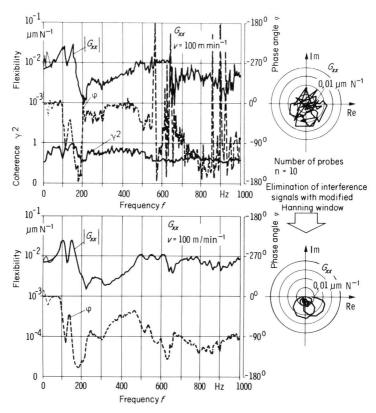

Fig. 4.41 Interference signal elimination with a modified Hanning window

4.7 Analysis of vibration form or mode

4.7.1 Mode analysis of vibrations under sinusoidal input signals

When polar curves for the flexibility of machines are plotted it is usual to watch for any marked increase in the resonance of the flexibility. In order to establish which machine components make a major contribution to the resonance phenomenon, it is necessary to determine the form or mode of the vibration. Figure 4.42 is a schematic diagram of the equipment necessary for such an investigation using the spindle/bearing system of a lathe as an example.

The machine is excited with a sinusoidal test force signal into the resonance frequency which is of interest. The vibration amplitudes are measured at various points of the object under test with transducer 2 (displacement, velocity or acceleration transducer). By comparing the phase relationship of the signal of transducer 2 with that of the fixed transducer 1, the direction of the vibration is established. As only the signs of the respective phases are of interest, i.e. in phase or out of phase, this information is most easily obtainable with the aid of Lissajous figures.

By taking a systematic series of readings on the machine this method will identify the dominant vibrating machine components and establish their

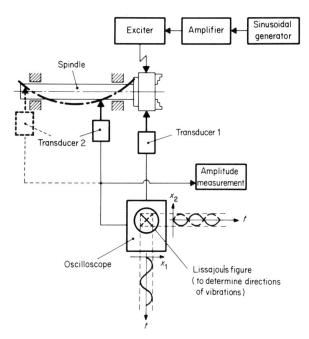

Fig. 4.42   Equipment arrangement for determining
the form or mode of vibrations

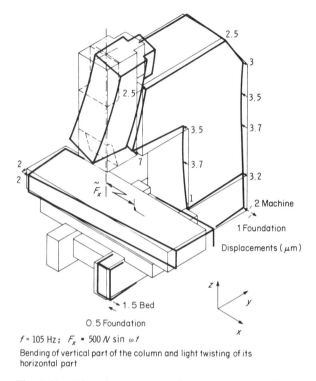

f = 105 Hz; F_x = 500 N sin ω t
Bending of vertical part of the column and light twisting of its
horizontal part

Fig. 4.43   Vibration   mode   of   a   bed-type   milling
machine

dynamically weak points. As a result, it becomes possible to take appropriate steps to improve the dynamic behaviour.

Figure 4.43 shows the vibration mode of a bed-type milling machine as a result of alternating force excitations in direction x at a resonance frequency $f_R = 105$ Hz. The bending of the vertical part of the column and the small twist of its horizontal section make a major contribution to the displacement at the cutting point under this type of excitation.

4.7.2   Mode analysis of vibrations under stochastic and aperiodic input signals

To determine the weak points the establishment of vibration modes for individual machine resonance positions can also be achieved with stochastic or aperiodic test-force signals. In this case, the transmission functions must be established in three co-ordinates for a given dynamic loading condition at the individual displacement investigation points on the machine contour. By evaluating the measured flexibility amplitudes for the individual resonance positions, it is possible to determine the vibration vectors with respect to their three-dimensional and time positions at each point under investigation, pro-

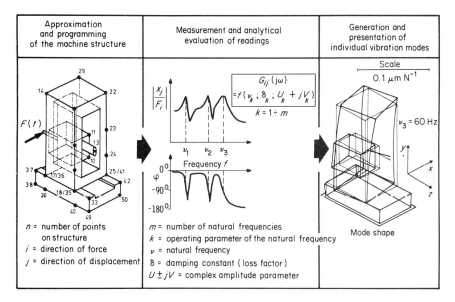

Approximation and programming of the machine structure	Measurement and analytical evaluation of readings	Generation and presentation of individual vibration modes

n = number of points on structure
i = direction of force
j = direction of displacement

m = number of natural frequencies
k = operating parameter of the natural frequency
v = natural frequency
δ = damping constant (loss factor)
$U \pm jV$ = complex amplitude parameter

Fig. 4.44 Outline of procedure for a Modal analysis

viding the appropriate phase relationship has been taken into account. This enables a complete presentation of the vibration modes to be made.

An automatic evaluation and presentation of the results can be made with the so-called Modal analysis[21] which is computer-assisted. In addition, the vibration mode may also be presented on the screen of a fast-Fourier analyser as a vibrating machine structure. The basic procedure for a Modal analysis is shown in Fig. 4.44. The main steps are as follows:

(a) the geometric approximation and data input for the machine structure;
(b) measurement and analytical evaluation of the three dynamic flexibility – frequency response characteristics in respect of each point under test; and
(c) the generation and presentation of the individual vibration modes.

The decisive mathematical operation in a Modal analysis is the approximation of the measured transmission functions by an analytical system of equations; this consists mainly of a sum of balanced terms for each important natural frequency (a definition of the nomenclature is in Fig. 4.44):

$$G_{ij}(j\omega) = -\frac{1}{m'_{ij}\omega^2} + \sum_{k=1}^{m}\left[\frac{U_{ijk} + jV_{ijk}}{-\delta_k + j(\omega - v_k)} + \frac{U_{ijk} - jV_{ijk}}{-\delta_k + j(\omega + v_k)}\right] + S''_{ij} \quad (4.13)$$

Each natural frequency (k) of the machine structure under investigation is then defined by four so-called 'Modal parameters'. The damped natural frequency v_k and the damping constant (or loss factor) δ_k are fixed magnitudes for the structure. They are identical at all points of the structure for a resonance value k. In contrast, the complex amplitude parameter $U_{ij} \pm j\,V_{ij}$ varies for each point on the structure in accordance with the flexibility condition. The residual terms $1/m'_{ij}\omega^2$ and S''_{ij} take into account the influence of the natural frequencies below or above the approximated frequency range.

The Modal parameters and both residual terms are determined using a repetitive calculation procedure by the 'least-squares' method between the measured and analytically approximated function.

4.8 Measures for reducing chatter tendencies

It is often necessary in practice to take positive steps to avoid chatter vibrations. A discussion of the various influences of a machining process upon stability conditions provides the opportunity to summarize the various measures that may possibly be taken which will tend to have a stabilizing effect (Fig. 4.45). In order to provide a cross-reference to Fig. 4.13, the classification headings for the applications are machine, workpiece/tool and cutting process.

When choosing the most suitable measures, consideration must be given to the practicality for given cases. For machines which are already in service usually only those measures are possible which are intended to change the directional orientation, i.e. the type of stress, and the cutting process condi-

	Direction of application			
	Machine		Workpiece/tool	Cutting process
	Operating conditions	Directional orientation		
Measures	1. Increase in static stiffness 2. Stiff foundations or well damped mounting 3. Choosing optimum positioning of machine components (slides, cross beams, tool-holder positions) 4. Variation of speeds to minimize regenerative effect 5. Exploitation of non-linear effects 6. Improvement of damping of systems: active damper, passive damper, damping support, damping strip	Choice of machining condition so that the: 1. resultant cutting force or 2. normal line to the cutting face is perpendicular to the direction of the highest dynamic flexibility of the machine	1. Support of resilient workpieces (steady) 2. Low mass workpiece 3. Rigid mounting of workpiece 4. Effectively damped tool 5. Low mass of tool	1. Choice of work material for low k_{cb} values 2. Minimizing clearance angles 3. Negative rake 4. Stoning of cutting edge 5. Increase of feed rate 6. Choice of low or very high cutting speeds to avoid minimum stability point 7. On multiple cutting-edge tools, choice of uneven number of cutting edges 8. Choice of stable spindle rotation (between chatter peaks)

Fig. 4.45   Measures for reducing chatter tendencies

Description	Application examples	Functional principles
Passive — energy absorbing		
Friction damping (dry friction)	Frame components, fitting of friction plates, leaving sand in cast components	
Auxiliary mass damper (Lanchester damp) (impact vibration damper)	Machine columns and tables; boring bars; lathe, milling and grinding machine spindles, rams, lathe tools	
Adjustable auxiliary mass damper	Milling machine and lathe spindles, rams	
Damping sleeve (squeeze-film effect)	Lathe, milling and grinding machine spindles	
Active — energy producing		
Active damper	Machine tables, tool slide, cross beams resilient workpieces	
Controlled mechanical impedance	Tool holder	

Fig. 4.46 Active and passive damping systems

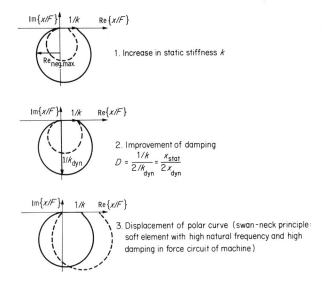

1. Increase in static stiffness k

2. Improvement of damping
$$D = \frac{1/k}{2/k_{dyn}} = \frac{x_{stat}}{2x_{dyn}}$$

3. Displacement of polar curve (swan-neck principle: soft element with high natural frequency and high damping in force circuit of machine)

Fundamentals: reduction of the maximum negative real position of the flexibility polar curve

Fig. 4.47 Measures to reduce chatter tendency

tions. In addition, the dynamic machine behaviour can often be improved by an additional increase in the system damping by means of active or passive dampers.

The measures indicated in Fig. 4.46 are under no circumstances to be considered merely as remedial steps; on the contrary, they are widely used and constitute very effective design features which will improve dynamic behaviour. Their function is to increase the overall damping of the system.

In Fig. 4.47, it may be seen how three basic measures for improving chatter stability affect the flexibility polar curves. The aim of all measures is to reduce the maximum negative real component of the polar curve, which is a direct measure of the chatter tendency of the machine.

4.9 Evaluation of flexibility conditions; machining simulation

As has been mentioned earlier, a problem arises when determining machine characteristics by direct measurement because the measured quantity (e.g. in this case the static and dynamic flexibility) cannot be simply interpreted with regard to its influence upon the machining conditions. Hence it is necessary to simulate the relevant conditions of loading which will produce readings from which the cutting conditions can be described. By applying equation (4.12) from section 4.3 it is possible to determine the limit of stability in terms of the limiting critical chip width in relation to material and machine parameters. The cutting-process variables are kept absolutely constant, so that an objective evaluation of the machine is made possible.

As will be seen in the following paragraphs, a major advantage of this process is derived from the fact that a single determination of a so-called flexibility matrix by a geometric transformation processed by a digital computer enables any desired loading conditions to be simulated.

A general description of the relative flexibility behaviour at the cutting point is necessary with respect to each measurement point on the machine under test. As the force excitation in any given direction produces a displacement not only along the same axis but also usually in the direction of the other two co-ordinates as well, the flexibility conditions at the cutting point may be defined with the following system of equations:

$$\underbrace{\begin{Bmatrix} x \\ y \\ z \end{Bmatrix}}_{\{X\}} = \underbrace{\begin{bmatrix} G_{xx} & G_{yx} & G_{zx} \\ G_{xy} & G_{yy} & G_{zy} \\ G_{xz} & G_{yz} & G_{zz} \end{bmatrix}}_{[G]} \underbrace{\begin{Bmatrix} F_x \\ F_y \\ F_z \end{Bmatrix}}_{\{F\}} \qquad (4.14)$$

where $\{X\}$ = the displacement vector for the translatory magnitudes of movement in the corresponding axes x, y, z
$\{F\}$ = the stress vector for the forces F_x, F_y, F_z, along axes x, y, z
$[G]$ = the flexibility–frequency matrix consisting of nine individual frequency characteristics of which three are

direct frequency characteristics along the main diagonals and six are cross-frequency characteristics. The first suffix of the frequency characteristics defines the vector direction of the source, i.e. the forces in the x, y and z directions; the second suffix identifies the direction of the reaction in terms of translatory movement. For linear systems such as those normally encountered on machine tools $[G]$ is symmetric, i.e. $G_{xy} \equiv G_{yx}$, $G_{xz} \equiv G_{zx}$ and $G_{yz} \equiv G_{zy}$.

For an accurate determination of the flexibility matrix $[G]$ the forces F_x and F_y are not applied at the cutting point but are directed in such a manner that their resultant direction passes through the spindle axis. The rotary transmission behaviour and the consequential introduction of torques in the machine structure are therefore ignored. This simplification is permissible, as experience has shown that changes in chip thickness due to variations in torque are normally insignificant.

Figure 4.48 shows a flexibility matrix for one test position on a bed-type milling machine. Based upon this generalized concept for defining the flexibility behaviour at the cutting point, the directional flexibility – frequency characteristic $G_d(j\omega)$ (see section 4.3) may be expressed for the general case as:

$$G_d(j\omega) = \sum_i \sum_j d_{ij} G_{ij}(j\omega) \quad \text{for } i, j = x, y, z \tag{4.15}$$

By the multiple linking of so-called 'direction factors' with the corresponding orthogonal flexibility – frequency response curves, it is possible to calculate the directional transmission frequency characteristics from a single reading of the frequency matrix in the machine co-ordinates for any machining condition

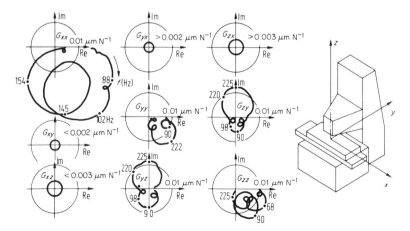

Fig. 4.48 Flexibility matrix for one test point on a bed-type milling machine

92

which may be required. The direction factors enable the general three-dimensional force to be projected into the directions of the co-ordinates x, y and z, and similarly the resultant displacement into the direction of the change in chip thickness, i.e. the direction normal to the cutting surface.

The direction factors take the geometric conditions for a given machining condition into account, i.e. the position of the tool relative to the work, the cutting-tool geometry as well as the number of cutting edges on the tool employed.[13] The relationship given in equation (4.12) may be diagrammatically presented in a block diagram (Fig. 4.49). In the case of a simple tool/work configuration the directional flexibility – frequency response characteristic may be obtained by an appropriate arrangement of the exciter and transducer.

Figure 4.50 shows in diagrammatic form the set-up for obtaining the directional polar curve for a lathe. Such a test set-up may be used for establishing the directional polar curve for the 'plunge cutting ($\beta = \beta_{\text{orth}}$)' condition which

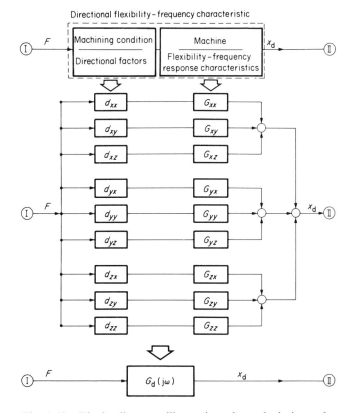

Fig. 4.49 Block diagram illustrating the calculation of a directional flexibility–frequency response characteristic taking the three translatory degrees of freedom into account

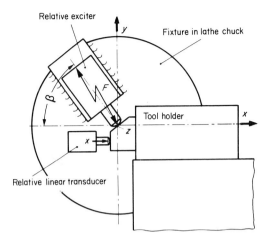

Fig. 4.50 Testing set-up to determine the
directional polar curve on a lathe

is usually crucial for a lathe. The angle β is that which is enclosed between the directions of the chip-thickness change and the dynamic cutting force and is governed by the cutting geometry and the material being cut. Therefore, a number of directional polar curves must be produced which embrace the likely range of the angle β. (When cutting with a conventional cutting edge β_{orth} will be between $76°$ and $89°$; for grinding operations $\beta_{orth} = f$ (in-feed) $<30°$.) For this reason, it is recommended that the direct frequency characteristic in the direction x, i.e. $G_{xx}(j\omega)$, is measured, as well as the cross-frequency characteristic $G_{yx}(j\omega)$, so that the directional polar curve for any given angle may be determined from the relationship:

$$G_d(j\omega) = G_{xx}(j\omega) \cos \beta + G_{xy}(j\omega) \sin \beta \qquad (4.16)$$

With the aid of equation (4.16) the directional flexibility – frequency characteristics may be calculated for a large number of lathes and the critical limiting chip width may be determined based on equation (4.12). At the same time, the chatter threshold during the cutting tests may be established for comparable boundary conditions (see the order of testing procedure in Chapter 7).

In order to establish the linear relationship between the theoretically calculated dynamic characteristic magnitudes and the critical limiting chip widths obtained during the chatter test, the ratios of the latter values ($b_{cr\ trial}$) to the former ($b_{cr\ calculated}$) are calculated (see Fig. 4.51). As may be noted from the diagram, the relationship between the limiting chip widths obtained by the cutting test and those calculated on the basis of the flexibility measurements is not strictly proportional. The spread of the individual values which is in the form of a gaussian distribution with a mean value of 1, even for such a relatively small number of test results, indicates that a definite systematic

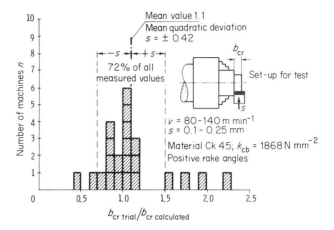

Fig. 4.51   Relationship between actual and theoretically calculated critically limiting chip widths

relationship exists between the parameters under examination. At the same time a comparatively wide scatter band is observed.

The reason for this wide scatter may be traced to the various influences upon chatter which are not precisely determinable (tool condition, geometry, work material, etc.) and emphasizes the uncertainty of the results obtained from machining tests. A further complication is introduced in the case of milling machines due to the need for examining a large number of tool/work-piece configurations, i.e. the effect of the position and length of the cutter arc on each point of measurement.

Due to the changing tool/workpiece configuration the machine is subjected to varying loading conditions which result in differing limiting chip-width values when considered in conjunction with the overall flexibility behaviour of the machine. This test condition is represented in the face-milling operation shown in Fig. 4.52. This face-milling simulation may be carried out for a range of approach angles and cutting arcs which will encompass all the relevant tool/workpiece configurations and cutter geometries which may be expected in practice.

If one starts from the standpoint that the minimum limiting cutting depth of the stability chart is of interest because it is not possible to determine the stable ranges between the U-shaped curves without knowledge of the stability chart, then it is logical to evaluate this minimum limiting cutting depth for a given machining condition for all positions of the workpiece around the tool. Figure 4.53 shows the results of such a machining simulation for a bed-type milling machine, the flexibility matrix of which is given in Fig. 4.48.

The minimum limiting cutting depth is shown in relation to the angular position of the cutting arc ϕ_m in two polar diagrams for the approach angles $\kappa_M = 90°$ and $\kappa_M = 75°$ and for various magnitudes of the cutting arc, viz.

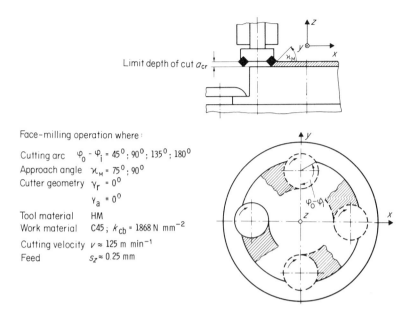

Limit depth of cut a_{cr}

Face-milling operation where :

Cutting arc	$\varphi_0 - \varphi_i = 45^0 ; 90^0 ; 135^0 ; 180^0$
Approach angle	$\varkappa_M = 75^0 ; 90^0$
Cutter geometry	$\gamma_r = 0^0$
	$\gamma_a = 0^0$
Tool material	HM
Work material	C45 ; $k_{cb} = 1868\ N\ mm^{-2}$
Cutting velocity	$v \approx 125\ m\ min^{-1}$
Feed	$s_z \approx 0.25\ mm$

Fig. 4.52 Representative operating conditions for a mathematical simulation to determine limiting cutting depth

$\phi_0 - \phi_i = 45°$, 90°, 135° and 180°. This provides a visual impression of the stability conditions of the machine in relation to the position of the cutting arc within the machine's co-ordinate system (for vertical machines the xy plane is coincident with the table plane).

The hyperbolic shape of the curves in the polar diagrams is typical for the representation of limiting cutting depths in relation to the positional angle of the cutting arc (direction orientated). At the instability points of the curves the source of the chatter changes. It is also typical that the limiting cutting depth and hence the minimum dynamic stiffness first decreases ($\phi_0 - \phi_i = 45°$ to 90°) as the cutting arc increases (which increases the tendency to chatter due to the larger number of cutting edges in contact with the work). When the cutting arc $\phi_0 - \phi_i$ is greater than 90° the limiting cutting depth tends to become larger.

The pattern of these curves shows that the stability behaviour of a machine varies considerably as the tool/workpiece configuration changes, even when the work produced is unchanged and therefore the cutting arc and the setting angle must be considered as important parameters. Computer programs are available to calculate stability charts and the lowest useful limiting cutting conditions, whilst also considering the direction-orientated limiting factors. The sequence of the program is largely automated and therefore the input data consist solely of the machine's flexibility polar curves and the parameters of the machining process which is to be simulated.

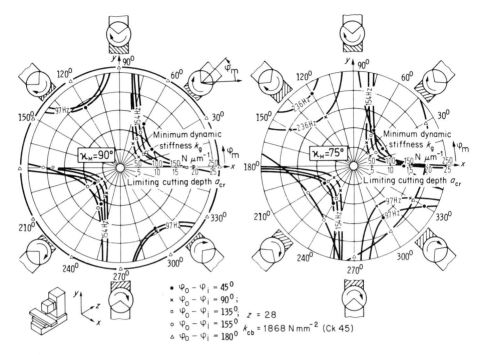

Fig. 4.53 Limiting depth of cut and minimum dynamic stiffness in relation to position and magnitude of the cutting arc and the approach angle κ_M

In this manner, an objective and comprehensive definition of the dynamic behaviour of metal-cutting machine tools becomes possible. However, as may be seen from Fig. 4.53, the number of parameters that must be considered when making comparative studies of machines is so large that it is necessary to markedly reduce the amount of the data by the use, for example, of statistical evaluation methods. Suggestions have been given for such a procedure with promising success.[13,22]

5

NOISE EMISSION BY MACHINE TOOLS

A large number of machine operators work less efficiently and even suffer physical harm in the long term as a result of excessive noise in the machine shop. Consequently, hearing defects are the most common of all occupational illnesses for which the majority of trade unions seek compensation for their members.

Two fundamental aims may be noted when measuring and evaluating noise. Noise-penetration investigations seek to determine the effect of noise on human beings by using suitable techniques. On the other hand, noise-emission studies are concerned with noise radiations from particular sources, e.g. machines, with the aim of producing comparative conclusions with regard to the acoustic conditions.

After explaining some basic concepts, the following sections will deal with measures available for noise appraisal and analysis. For the purpose of noise-emission examinations in detail, lathes are considered as typical examples.

5.1 Basic concepts, sound characteristics

Sound may be defined as mechanical vibrations of solid, fluid or gas substances at frequencies from about 16 Hz to a maximum of 16000 Hz. Within this range of frequencies the human ear will detect the vibrations as sound.[23-26]

The fundamental measuring unit is the sound pressure; it is a time-dependent pulsating pressure superimposed upon the atmospheric pressure (order of magnitude 10^2 to 10^{-5} Nm^{-2}). For the evaluation of sound characteristics, the effective value is required to be determined from the sound pressure which fluctuates with respect to time:

$$\tilde{p} = \sqrt{\left[\frac{1}{T}\int_0^T p^2(t)\,\mathrm{d}t\right]} \qquad \text{where } T \to \infty \qquad (5.1)$$

Technical measuring devices determine the effective value during a finite time T (order of magnitude 1 s).

Acoustic velocity is defined as the velocity at which the constituent particles oscillate in the sound field. The effective value is calculated from:

$$\tilde{v} = \sqrt{\left[\frac{1}{T}\int_0^T v^2(t)\,dt\right]} \qquad \text{where } T \to \infty \tag{5.2}$$

This value must not be confused with the velocity of sound c (velocity of propagation or radiation of the sound energy). The acoustic velocity is generally proportional to the sound pressure p.

The sound intensity is the sound energy which passes through the unit area in unit time:

$$I = \tilde{p}\tilde{v} \qquad \text{N m}^{-1}\text{s}^{-1} \tag{5.3}$$

When $\tilde{v} = \tilde{p}/\rho c$ we have: $\tag{5.4}$

$$I = \frac{\tilde{p}^2}{\rho c} \qquad \text{W m}^{-2} \tag{5.5}$$

where ρc is the characteristic sound impedance (408 N s m^{-3} for air), ρ being the density of the medium and c the velocity of the sound.

The acoustic power is obtained from the sound intensity and the actual area S through which the sound passes:

$$P = IS = \tilde{p}\tilde{v}S = \frac{\tilde{p}^2}{\rho c}S \qquad \text{W} \tag{5.6}$$

For the total radiated acoustic power of a sound source we have:

$$P = \phi \, \mathbf{IdS} \tag{5.7}$$

5.1.1 Sound pressure level and acoustic power level

Sound pressure can be detected by man over a wide pressure range. At a frequency of 1000 Hz the threshold of audibility (i.e. the lowest level at which a sound pressure is detectable by human hearing) is at an effective value of 2×10^{-5} Nm^{-2} and the threshold of pain is at an effective sound pressure of 20 Nm^{-2}, as shown in Fig. 5.1. The pressure and frequency ranges applicable to speech and music are also shown in the diagram. Due to this very large range, it is not very meaningful to give a definition of sound by stating the sound pressure value; moreover, a linear scale of the sound field magnitudes has little in common with human sensitivity to sound.

These factors have led to the definition of sound level units for use in technical acoustic studies. The name of this dimensionless unit is 'decibel' (dB).

The sound pressure level L is:

$$L = 10 \lg \frac{I}{I_0} \tag{5.8}$$

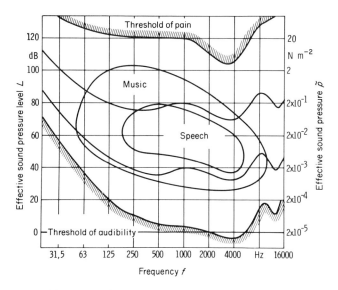

Fig. 5.1   Diagram of hearing ranges (L. Cremer)

When $I = \tilde{p}^2/\rho c$ and $I_0 = 1 \times 10^{-10}\,\mu\text{W cm}^{-2}$ we have:

$$L = 10 \lg \frac{\tilde{p}^2}{p_0^2} = 20 \lg \frac{\tilde{p}}{\tilde{p}_0} \tag{5.9}$$

The value of $\tilde{p}_0 = 2 \times 10^{-5}\,\text{N m}^{-2}$ is taken as the reference sound pressure.

Term	Nomen-clature	Formula
Effective sound pressure	$\tilde{p}$	$\sqrt{\left[\frac{1}{T}\int_0^T p(t)^2\,dt\right]}$; $T \to \infty$
Effective acoustic velocity	$\tilde{v}$	$\sqrt{\left[\frac{1}{T}\int_0^T v(t)^2\,dt\right]}$; $T \to \infty$
Sound intensity	I	$\tilde{p}\,\tilde{v} = \frac{\tilde{p}^2}{\rho c}$
Acoustic power	P	$\oint I\ dS$
Sound pressure level	L	$10 \lg \frac{I}{I_0} = 10 \lg \left(\frac{\tilde{p}}{\tilde{p}_0}\right)^2$
Acoustic power level	L_W	$10 \lg \left(\frac{\tilde{p}}{\tilde{p}_0}\right)^2 + 10 \lg \frac{S}{S_0}$ $(L + L_S)$
Reference values $I_0 = 10^{-10}\,\mu\text{W cm}^{-2}$; $\tilde{p}_0 = 2\times10^{-5}\,\text{N m}^{-2}$; $s_0 = 1\,\text{m}^2$		

Fig. 5.2   Basic concepts in sound technology

The acoustic power level L_w is:

$$L_w = 10 \lg \frac{P}{P_0} \qquad (5.10)$$

When $P = \bar{p}\bar{v}S$ and with the reference sound pressure P_0 we have:

$$L_w = 10 \lg \frac{\bar{p}^2}{\bar{p}_0^2} + 10 \lg \frac{S}{S_0} \qquad \text{where } S_0 = 1 \text{ m}^2 \qquad (5.11)$$

The acoustic power level is therefore derived from the sound pressure level and an additional term which takes the measured surface into account.

Figure 5.2 is a summary presentation of the basic concepts necessary for understanding and evaluating sound problems in the practice of noise measurement.

5.2 Evaluation of sound levels

Normal human hearing can detect frequencies of approximately 16 Hz to 16 kHz and the ear is most sensitive to sounds in the middle of this range. The relationship of the level of loudness to the sound pressure level and frequency is described by 'curves of equal loudness' (isophones), as shown in Fig. 5.3. These curves indicate the level at which a sound signal at a given frequency must be produced, so that it appears just as loud as a 1000 Hz signal with a given sound pressure level.

A pure tone at a frequency of 50 Hz and sound pressure level of 80 dB will

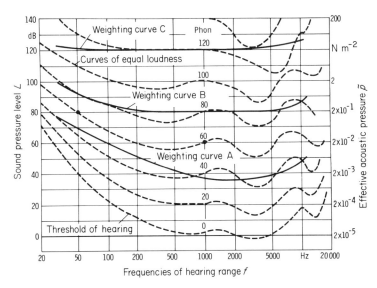

Fig. 5.3   Curves of equal loudness for sinusoidal frontal incident single tones heard with both ears

seem just as loud as the standard sound of 1 kHz and 60 dB, i.e. the tone has a loudness level of 60 phon.

In order that the frequency-dependent sensitivity of the ear is taken into account when noise sources are scientifically examined, a frequency evaluation has been established which weights the sound in relation to its frequency. Weighting curves A, B and C have been produced (solid lines in Fig. 5.3). The A curve is an approximation of the curves of equal loudness valid for acoustic power levels less than 55 dB, the B curve for values between 55 and 85 dB and the C curve for acoustic power levels over 85 dB. International agreements, however, specify that the A curve is used today for all acoustic power levels when a noise is evaluated in relation to its frequency, so that comparable results are obtained. As these evaluation curves represent an approximation of the isophones, the results may not be expressed in terms of 'phon'. The proper nomenclature for the weighted noise level is dB(A).[27]

5.3 Evaluation of time-dependent decaying noise

5.3.1 Permissible effective times for sound pressure level

The main purpose of all noise investigations is to minimize the effect of the noise upon people, so that they are protected from dangers and possible damage to their health. After extensive medical research limiting values have been established which must not be exceeded.

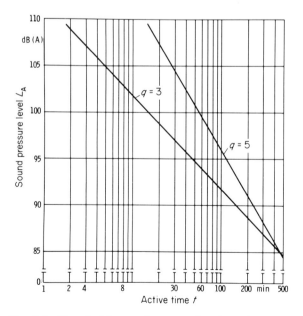

Fig. 5.4 Permissible sound pressure level as a function of its effective time for $q = 3$ and $q = 5$ (UVV noise)

Figure 5.4 shows the permissible sound pressure level in relation to its duration. For an eight-hour working day (480 min) the maximum permissible value is 85 dB(A)[28] (90 dB(A) in the United Kingdom). If this limit is exceeded, then in accordance with the UVV Noise standard (Accident Prevention Order) the management is required to provide personal ear defenders. It is clear from the graph that for a change in the level of 3 dB(A) the permissible effective time varies by a factor of 2.

5.3.2 Intermittent noise patterns—equivalent level

In general, noise phenomena do not occur continuously but change with respect to time. Consequently, in accordance with DIN 45641[29] a mean-time level L_m, the energy of which is equivalent to a continuous noise (energy equivalent level), is calculated within suitably chosen time divisions. The time divisions t_i are chosen in such a way that similar sound actions are combined. In particular, continuous and periodic noise phenomena are easily segregated into time divisions. The measurement time within these time divisions is governed by the consistency of the sound pressure level. As may be seen in Fig. 5.5, a short measuring time is adequate in time division 1 because the sound pressure level is almost constant. In the time division 2 at least one complete cycle must be encompassed. The measuring time in the third case is identical with time division 3 because the sound pressure level is completely irregular during this time:

$$L_m = 10 \lg\left(\frac{1}{T} \sum_{i=1}^{n} 10^{L_i/10} t_i\right) \qquad (5.12)$$

By the application of this calculation procedure an energy mean value is obtained from varying noise levels and their effective time periods, which may be compared with prescribed limiting values.

The evaluation level L_r is a modified time-dependent mean level. It is applied as a basis for evaluating noises with respect to their effect upon people over a longer time (evaluation time span), e.g. a shift, a day or a week:

$$L_r = 10 \lg\left(\frac{1}{T_r} \sum_{i=1}^{n} 10^{(L_{mi}+K_i)/10} T_i\right) \qquad (5.13)$$

The evaluation level is calculated using the same type of formula as the mean level. However, the sound pressure level L_{mi} may have an added term K_i to take the tone and pulse time or particular times of the day or night into account insofar as these are not considered in the measured data. The individual noise situation is considered by means of level additions in the evaluation level. The magnitude of the additions is laid down, for example, in the standard VDI 2058 sheet 1.[30]

Another simple method for evaluating time-decaying sound is the cyclic

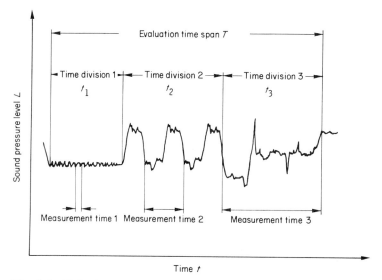

Fig. 5.5   Noise-level pattern with evaluation time span and examples
of suitably chosen measurement and time divisions

maximum level technique.[31] Within constant time divisions, the number of
times that the maximum sound pressure level is reached is recorded and the
mean level is then calculated:[29]

$$L_{mT} = 10 \lg\left(\frac{1}{H} \sum_{i=1}^{n} 10^{L_i/10} h_i\right) \qquad (5.14)$$

5.4   Spectral composition of sound

Apart from the conclusions which may be reached from the system of sound
measurements described with respect to the design and layout to produce
machines with a low noise output, special importance may be attached to the
systematic analysis of noise sources. To determine the cause of the noise
being produced by individual sources, it is necessary to make a frequency
analysis of the air-borne and structure-borne sounds. Let us first briefly consider the relationship of signals to the time and frequency ranges.

In Fig. 5.6 an arbitrary measuring signal is shown on a time base. Every
signal can be resolved into individual harmonic components of different amplitudes and phase relationships by the use of a Fourier transformation or by
the use of filters.[20] In this example the output signal consists of three sinusoidal vibrations of 1000, 2000 and 3000 Hz respectively. In the lower part of
the diagram information is clearly presented about the individual frequency
components. The various phase relationships have not been ascertained.

A classification may be made of various 'characteristic frequency spectra'.
In Fig. 5.7 the sound pressure amplitudes are shown on a frequency base for

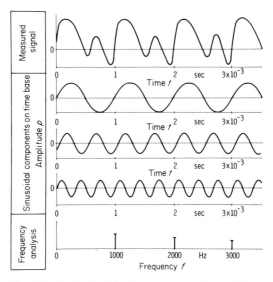

Fig. 5.6 Relationship between signals and time
and frequency ranges

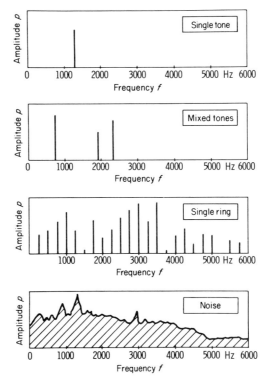

Fig. 5.7 Amplitude spectrum for various types
of sound

various types of sound:

Single tone: sinusoidal sound
Mixed tones: sound composed of tones of differing frequencies
Single ring: sound composed of harmonic tone parts
Noise: mixed tones which follow a continuous spectrum or are composed of a large number of single tones whose frequencies are not in an integer relationship to each other

5.5   Sound measurement techniques

5.5.1   Measurement of sound level

The sound which is radiated from a noise source is absorbed by a microphone as a sound pressure converted into an electrical signal and amplified (Fig. 5.8). The measured sound pressure is corrected in accordance with the weighting curve A (see Fig. 5.3) and thus interpreted in terms of the sensitivity of the human ear. An electronic integration circuit produces the effective value from the evaluated sound pressure signal and indicates the result as a sound-level magnitude. DIN 45633[27] specifies that the integration is carried out in three stages for precision sound-level meters. Consequently a time-based evaluation of the measured sound signal is obtained which simulates the physiological mechanism of hearing and takes this into account in the value indicated. In particular, for pulsating noise conditions, the indicator can be set to 'pulse' to produce a 'hearing-corrected' measurement of the sound.

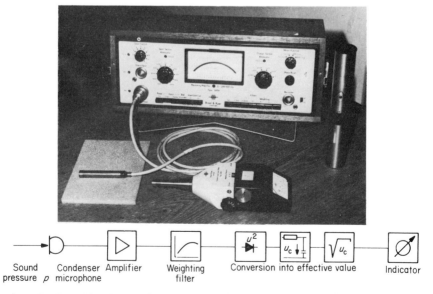

Fig. 5.8   Sound level meter in accordance with DIN 45633 for noise measurement

106

Two typical commercial measuring devices are also shown in Fig. 5.8. The unit in the lower part of the picture is a hand-held sound-level meter which is fitted with A, B and C weighting filters. Two different integration times may be chosen. The sound-level meter in the upper part of the picture can have an additional external filter for frequency analysis connected to it.

The determination of the mean or evaluation sound level may become very costly, depending upon the time that the noise is in existence. Commercially available automatic measurement and analysing devices therefore ease the evaluation of time-decaying noises considerably.

5.5.2 Measuring techniques for analysis

Frequency analysis of noise may be classified into analysis with absolute bandwidth (in general, a narrow-band filter) on the one hand and analysis with relative bandwidth (e.g. a third-octave or octave filter) on the other. The band-pass widths of filters are as follows:

Narrow-band filters $f_u - f_l =$ constant   (e.g. 15 Hz)

Third-octave filter $f_u = \sqrt[3]{2}\, f_l$
Octave filter $f_u = 2f_l$
Mean frequency $f_m = \sqrt{(f_l f_u)}$ (geometric mean)

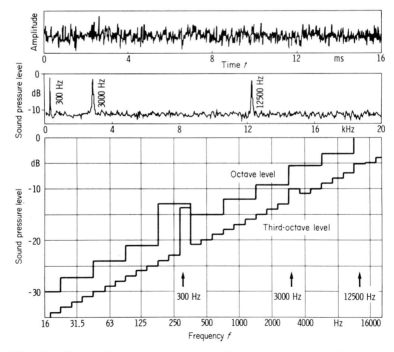

Fig. 5.9 Time–behaviour spectrum with constant bandwidth and third-octave and octave spectrum of sounds with separate tones

From the above, it may be seen that an octave filter encompasses three third-octave bands. DIN 45651 and DIN 45652 specify the nominal band-pass widths and mean frequencies of octave and third-octave filters.[32,33]

The various methods of operation and applications of different types of filters are indicated below.

In the upper part of Fig. 5.9 a sound is presented diagrammatically on a time-base consisting of three separate tone components (sinusoidal vibra-tions). Although the level of the components is 10 dB greater than that of the sound, they cannot be identified in the trace. The central diagram shows a relative bandwidth frequency analysis on which the individual tones may be clearly identified from within the basic sound. A frequency analysis with relative bandwidths (third-octave and octave) is then produced from the same time signal and the level of the mean frequencies is shown on a logarithmic frequency scale. This produces a step-shaped rising sound-pressure level curve. Moreover, it may be noted from the lower diagram that the identifica-tion of single tones with filters of proportional bandwidth becomes more difficult with increasing frequencies.

Only narrow-band analyses are successful for the determination of noise sources which are related to single frequencies. Due to the relatively simple measuring techniques, third-octave and octave analyses are applied for design exercises and simple noise evaluations.

5.6 Conclusions with respect to the source of noise

The source of air or body sounds (sounds which radiate in a solid medium or on its surface) may be established from their frequency analyses. The inves-tigations of geared drive units have been chosen as examples. In the upper part of Fig. 5.10, the drive is presented in diagrammatic form. The table below the diagram shows the speeds and rotational frequencies of the indi-vidual shafts for a motor speed of 1480 rev min^{-1} and particular clutch engagements, as well as the tooth-engagement frequencies of the respective gears in accordance with their number of teeth ($z \times n$). By comparing the data calculated in this way with the values obtained from a frequency analysis of the noise from the drive a direct relationship between the noise and the gear producing the sound can be established.

Figure 5.11 shows the set-up for a narrow-band frequency analysis. The measured air and body sound signals are fed into the main amplifiers (see also Fig. 5.8) from which the sound level and acceleration values may be read. The amplified signal is fed to a fast-Fourier analyser through a variable band-pass filter which is necessary to eliminate the frequency components that lie out-side the range under consideration. The digital analyser produces the fre-quency analysis by a mathematical Fourier transformation. The result may be seen on a cathode-ray display unit and may be obtained in hard-copy form from a digital plotter. The units can be operated from a remote teleprinter or by an adjacent push-button console. (A photograph of a Fourier analyser and its peripheral units is shown in Fig. 4.36 in section 4.6.)

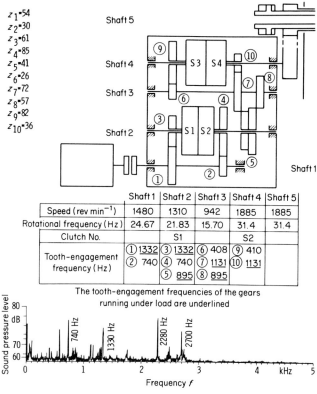

Fig. 5.10 Shaft speeds, tooth-engagement frequencies and spectrum of a drive

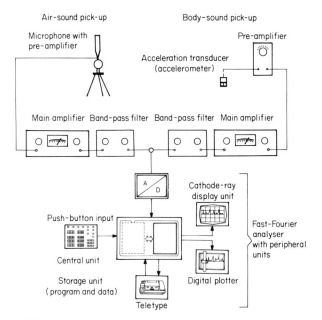

Fig. 5.11 Instrumentation for noise investigations

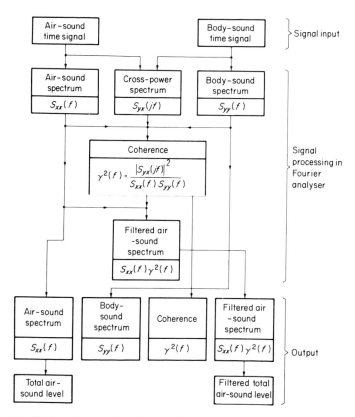

Fig. 5.12   Elimination of external noise interference with the use of the coherence technique

With a special program for the analyser it becomes possible to produce a frequency analysis when a simple analysis would lead to errors caused by significant external noise interference components. The main components of the so-called 'coherence technique' are shown in Fig. 5.12.[34]

The synchronized measured air and body sound signals are resolved into their spectral components. The coherence function is formed for both spectra. This establishes for each frequency the extent to which a causative relationship exists between the measured body sound and the air sound. Subsequently the frequencies which have a clear relationship to the body sound of the object under investigation as a result of good coherence are filtered from the total air sound spectrum. The amplitudes of the individual frequencies are added to a filtered total level which is free from external noise components (Fig. 5.13).

5.7   Measures for noise reduction

The relationships between noise emission and the measures which may be taken to achieve a reduction in the noise are shown in the block diagram in

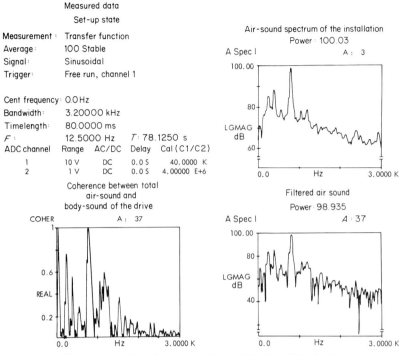

Fig. 5.13   Filtering of interference noise with the aid of the coherence function

Fig. 5.14. Active measures are those which are taken directly against the source, transmission or radiation of the sound from the vibrating machine elements. For certain types of machine or when the measures that are taken are remedial in nature often only passive steps will be effective. In this case, as in the case of active measures, a differentiation may be made between primary and secondary measures.

Figure 5.15 shows the floor plan of a lathe (motor, drive, working space, hydraulic unit and control cabinets) and the path along which the measurements were taken. The sound pressure level was measured in dB(A) at ten observation points. In order to obtain a reduction in the noise emission, the machine was covered with a sheet-metal casing which was left open at the top over the motor/drive space and over the hydraulic unit for cooling effects. Only the working space was totally enclosed. For a second investigation, the sheet-metal casing was replaced with a sound damping casing made from a sheet-metal/mineral wool/perforated sheet combination over the motor, drive and hydraulic unit and with a laminated sheet-metal case over the working space.

The results from the methods described above are presented in Fig. 5.16. The observations were taken while the machine was not cutting. The chain-dotted curve which closely follows the solid curve of the unenclosed machine

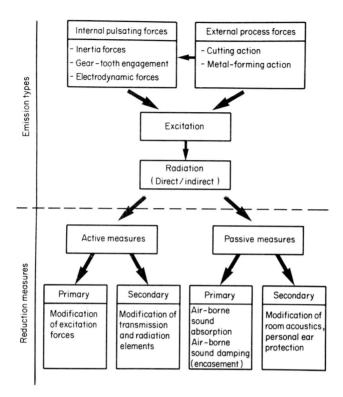

Fig. 5.14   Noise emissions and measures for their reduction

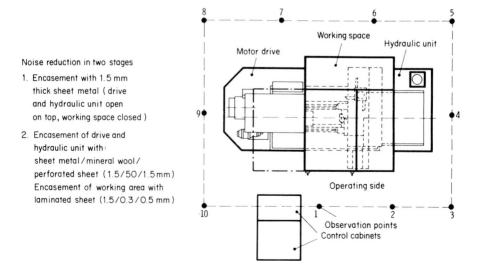

Fig. 5.15   Passive noise-reduction measures on a lathe (Gildemeister)

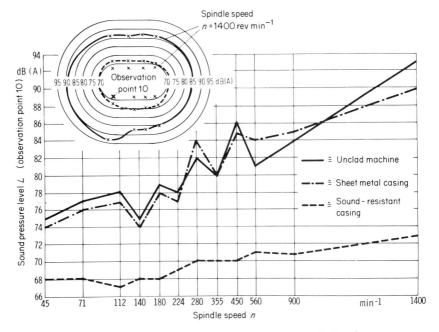

Fig. 5.16 Influence of sound-reduction measures on a lathe (measurement under no-load) (Source: Gildemeister)

noise emission indicates that the encasement with a simple 1.5 mm thick sheet-metal case produced no significant change in the noise emission. Only an enclosure using materials which have sound damping and sound-resistant characteristics (mineral wool, laminated sheet-metal) produces successful results. The dashed curve represents the improvement for observation point 10 in the speed-base graph as well as around the periphery of the machine for the maximum speed of 1400 rev min^{-1}.

Active noise-reduction measures are generally more effective when compared with passive measures, although they are not always more economic. Figure 5.17 shows an active measure taken for a single-stage spur-gear drive and the success achieved. The tooth depth of the gears was increased so that a near-integer profile overlap occurred under load. This produced a near-sinusoidal stiffness pattern without sharp changes and of low amplitude. A considerable reduction in sound pressure level was noted, particularly at high rotary speeds.

On blanking operations in press work, the cutting portion of the stroke or the shear-fracture length is the critical factor. The effect becomes more pronounced as the work material becomes more brittle. The shear-fracture length is determined by the elasticity of the press, tool and work material. The tool and press are pre-loaded during the cutting process with the force necessary to overcome the shear resistance of the work. This has the effect of

Gear-cutting data

m_n = 2.5 mm			a = 112.5 mm
z_1 = 36			z_2 = 54
d_{01} = 90 mm			d_{02} = 135 mm
α_{no} = 20°			β_0 = 0°
x_1 = 0.02075			x_2 = -0.02075
			b = 12 mm

	Normal spur teeth	Long addendum teeth
Tooth depth	5.625 mm	6.465 mm
Length of contact	12.759 mm	14.999 mm
Tool-head height factor	1.25	1.386
Profile overlap	1.729	2.032

Fig. 5.17 Reduction in noise radiation by modifying gear-tooth form

storing excess energy which is not used for the work process in the tool, press frame and drive. This spring energy is suddenly released at the point of shear fracture of the work material. The result is a free vibration of the force-transmitting elements, producing a strong sound radiation.

Hydraulic damping cylinders, as shown in Fig. 5.18, absorb and disperse in a controlled sequence the spring energy stored in the system upon its release after shear fracture. This markedly reduces the undesirable effects of the shear blow. The force – time diagram in Fig. 5.18 indicates the difference between a damped and undamped shear stroke.

The effect of shear-blow damping may be seen by the steadily declining and vibration-free force curve. The noise reduction is up to 12 dB(A) depending upon the method available for mounting the damping cylinders. The shear-blow damping shown here employs pre-loaded hydraulic cylinders. A common control valve is used for pre-loading both cylinders in accordance with the shear force required by the blanking operation. As the material fractures

114

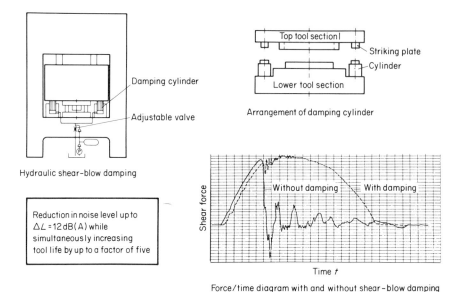

Hydraulic shear-blow damping

Arrangement of damping cylinder

Reduction in noise level up to
$\Delta L = 12\,dB(A)$ while
simultaneously increasing
tool life by up to a factor of five

Force/time diagram with and without shear-blow damping

Fig. 5.18   Sound reduction on a press by damping of shear blow (Schuler)

the inertia forces are absorbed by the damping cylinders. The fluid volume which is thus displaced is forced into the reservoir of the hydraulic unit.

After the ram has passed through its lower dead centre point, the depressed pistons of the damping cylinders are returned to their upper position by the reverse flow of the pressurized fluid and are again pre-loaded to the set value of the control valve. The nearer that the damping cylinders can be placed to the point of maximum spring in the tool, the more effective will be the shear-blow damping. Optimum results are obtained when the damping cylinders are mounted directly on the press tool.[35]

Another method for reducing noise on presses is shown in Fig. 5.19. On the left side of the diagram the force and ram displacement with respect to time are shown for shearing different materials. Due to the sudden drop of the force after fracture, vibrations with high amplitudes and variable frequency are detected in the force, as well as the displacement curves on the upper diagram indicating the generation of noise and vibrations in the machine mountings. In the lower diagram, the shear force decays at a steady rate. The shear blow has been eliminated and the operation proceeds for all practical purposes without noise and vibration. The press ram moves with near-constant velocity independent of the shearing operation.

The constructional design and control measures used to achieve the above are shown on the right side of the diagram. The ram movement is controlled by a double-acting hydraulic cylinder and monitored by a linear displacement measuring unit. The measured signal is compared with a displacement – time function which can be varied in accordance with the material being sheared.

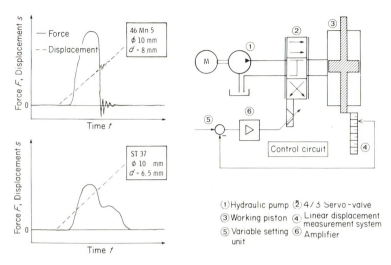

Fig. 5.19   Elimination of the shear-blow effect on presses by control of
the force with respect to time (Thyssen)

The amplified differential signal activates a 4/3 servo-valve which controls the working piston.

By eliminating the shear blow, noise-level reductions of up to approximately 30 dB(A) may be obtained. This will bring the noise from the press operation down to the range of noise which is found in general machine shops.

5.8   Evaluation of machine noise; current state of development of the technique

A statement with respect to the noise emission of a machine tool is of great interest to both the manufacturer and the user of the machine. The main indication of the extent of noise emission is the acoustic power level for given operating conditions. This may be used for comparative purposes for different machines with due consideration to external noise sources and allowing for acoustic reflections within the room in which the machine is situated.

The standard specification DIN 45635[4] provides a method of procedure. Figure 5.20 gives a diagrammatic presentation of the standard in three sections. The first is the 'scope section' in which generalized provisions may be found for the various measuring techniques. The second section contains the so-called 'procedural sheets'. In these there are instructions with respect to particular types of machines (e.g. machine tools, electrical machines, domestic machines). These are followed by appendices, containing specialized recommendations and operating conditions which must be applied when measurements are recorded for different machines.

116

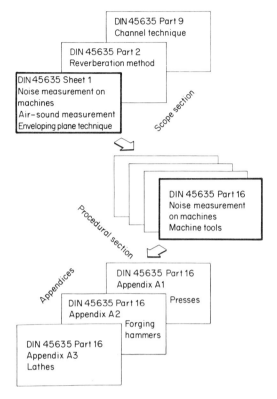

Fig. 5.20   DIN 45635—noise measurements of
machines

A simple and therefore practical technique for the determination of sound emission is the enveloping-plane technique. Enveloping planes are established at distances of 1 m around the geometrical idealized machine, as shown in Fig. 5.21. In accordance with the provisions of DIN 45635 Sheet 1, the microphone positions are established on these planes. The measuring points are arranged in such a way that the total sound effects of the machine may be established. The value of the term in equation (5.11) which takes the measured surface into account for the evaluation of the acoustic power level is obtained from the size of the measurement plane.

In a sound measurement, the sound energies of the machine under investigation are added to any other sound source which may be present in the test room. In order to be able to determine the noise emission which will permit an objective evaluation of the machine under test, the measured output levels must have contributions from any external noise removed from it. The magnitude of the correction value K_1 is dependent upon the difference between the total noise level and the extent of the external noise.

Moreover, the measured output is affected by acoustic reflections and diffractions in the test room. A simplified estimate of these ambient effects is

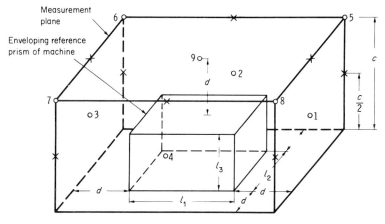

O Basic position of measuring points

X Additional measuring points when l_1 and or l_2 are greater than 1m

Fig. 5.21   Measurement plane, position and number of measuring points
(when $l_3 \leqslant 2m$)

Report for méasurement and evaluation of noise emission.

Investigation No. ...   Date : ....

　　　Machine ....   Type : ....

Measure-ment point No	Sound pressure level, dB(A) (average in unit time)	External sound level, dB(A) (average in unit time)	Correction value K_1, dB(A)	Corrected sound pressure level L_A, dB(A)
1				
2				
⋮				
n				
Mean sound pressure level　　　　　　=				
−Ambient effects K_2 = ... dB(A)				
= Sound pressure level of measuring plane $\bar{L}_A$　=				
+Measuring surface value L_S = ... dB(A)				
= A − Acoustic power level L_{WA}　=				

Sound pressure level at fixed operator's position L_A　　　=　...... dB(A)

Difference in level between 'spot pulse' measurements and 'slow' measurement at the loudest measuring point $L_{APm} - L_{ASm}$　=　...... dB(A)

Sketch 1 Floor plan of observation space; machine loading; position of machine under test	Sketch 2 Position and number of observation points on machine under test; enveloping planes	Note : Sketches should be provided on specially prepared sheets. If necessary, sound pressure spectra should also be provided on specially prepared sheets.

Fig. 5.22   Report for measurement and evaluation of noise
emission

made by the application of the correction value K_2. A prerequisite for this is knowledge of the ratio of the room volume V to the measuring area S and a classification of the ambient conditions into one of three categories. The correction value K_2 to eliminate the effect of the ambient conditions can then be obtained from the scope section of DIN 45635.

Figure 5.22 is a typical form which may be used to record the measurement and evaluation of a noise investigation. It shows the procedure for determining the acoustic power with allowances for external noise and the results of ambient effects. In addition, the sound pressure level at the fixed operator's position is to be stated (if available). This last factor is only necessary when pulsating noises occur in the test situation. A note is provided in the lower part of the form to the effect that the documentation should also include sketches of the observation space with the relative positions of the machine and the measuring planes, including information with respect to the number and positions of the microphones.

5.8.1 Description of the noise characteristics

A lathe is chosen to serve as an example for the determination of noise characteristics. The most important influences upon noise emission are the spindle speed and the instantaneous loading condition of the machine.

In a large number of consecutive tests, the acoustic power level of lathes of varying types was established with respect to torque and rotational speed. Figure 5.23 shows the observed sound-level pattern with respect to rotational speed n and torque T. The diagram also shows a balancing plane which was calculated by using the mathematical multiple regression technique defining the total characteristics and is used for simplifying the data. It can be seen from the law of the plane that if the rotational speed is doubled on this machine an increase of the sound level of 2.8 dB(A) will occur, whilst doubling of the torque will cause the level to rise by 1.6 dB(A).

The working range of the machine under test is restricted by the straight-line relationship of the nominal power (P_N = constant) in addition to the maximum rotational speed and the maximum torque. The correlation factor R and the residual variance s^2 provide quantitive assessments of the nearness of the mathematical approximation to the real measured values. The set-up necessary for such tests is shown in Fig. 5.24.[36]

The sound pressure level is measured at five points positioned on five enveloping planes around the machine and recorded on a magnetic tape together with the operating parameters which have been chosen. The measured data are prepared and processed in a digital computer. Peripheral units provide the output of the actual results and store the data for documentation and for further statistical evaluations.

The evaluation of the sound pattern shown in Fig. 5.23 is too costly for standard investigations; a representative loading case has therefore been developed for lathes and milling machines from a large series of tests which

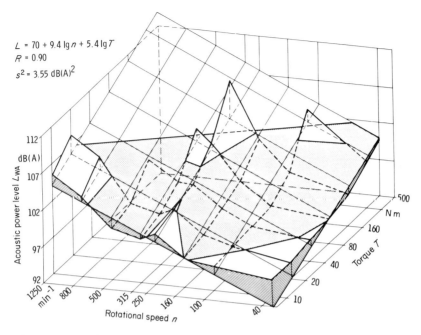

Fig. 5.23   Multiple regression diagram for the relationship between sound
levels and operating conditions

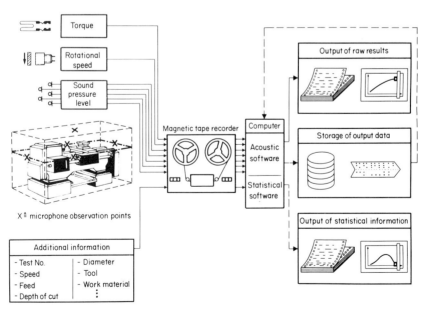

Fig. 5.24   Sound investigations on machine tools, presentation of development
to date in the technique

provides adequate information with respect to the sound-emission characteristics. The standard specification DIN 45635 defines the normal operating level for lathes at 50% of full load ($P = \frac{1}{2}P_N$; $n = \frac{1}{2}n_{max}$).

5.8.2 Current state of the technology

As a result of a multitude of machine investigations, as described above, average sound-emission characteristics of machine tools can be shown graphically in relation to their nominal power and maximum speed.[37] Such a presentation enables limiting values to be established which make it possible to classify the evaluation of sound emissions of machine tools and to state if the respective machines should be regarded as good, average or poor, in relation to the current state of the technology.

Figure 5.25 shows some limiting curves for lathes. Such a diagram can be used as a working document by an experienced engineer for evaluating the noise emission of a lathe. Due to the two variables of the machine, i.e.

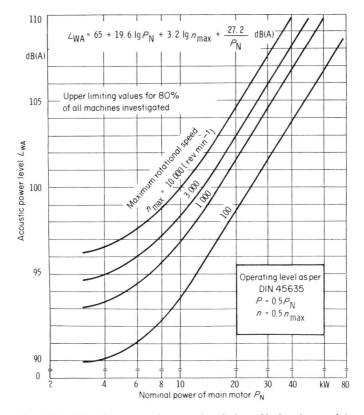

Fig. 5.25   Limiting curves for sound emission of lathes (example)

nominal power of the main motor P_N and maximum rotational speed n_{max}, a sound-emission value may be obtained from the diagram which will not be exceeded by 80% of all machines of this type currently in service when they are operated at 50% of full load, as specified.

INDIRECT MEASUREMENT OF THE EVALUATION PARAMETERS BY PERFORMANCE TESTS

6

WORK ACCURACY (EXAMINATION OF MACHINED WORKPIECES)

The quality of the components produced on a machine tool is a measure of the accuracy or degree of instability of the machine itself. Performance tests are undertaken by machining a number of similar workpieces to a prescribed design from a given material and subsequently measuring them. With the application of statistical evaluation techniques it is possible to separate systematic errors from random effects, as shown in Fig. 6.1.[3] Generally it is not possible to immediately identify the individual causes of work inaccuracy which are machine-orientated as shown along the bottom line of the block diagram because such influences have a collective effect upon the workpiece. Testpieces can, however, be designed in such a way that some individual

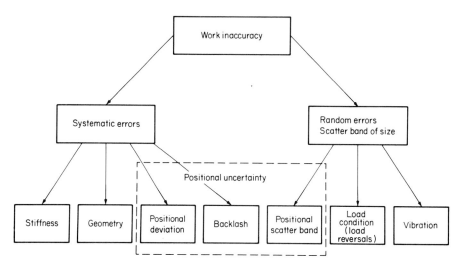

Fig. 6.1 Work inaccuracies of a machine—influencing factors according to VDI/DGQ 3441

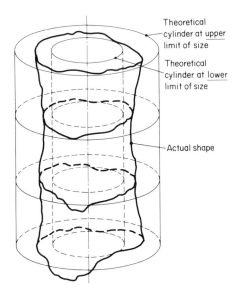

Theoretical cylinder at upper limit of size

Theoretical cylinder at lower limit of size

Actual shape

Fig. 6.2   Deviations in the form of a cylindrical component

influences upon the degree of overall inaccuracy may be separated (e.g. stiffness, geometry, positional accuracies). In addition to the separate influences which cannot be clearly identified but have a collective effect upon the work inaccuracy, the machining process is also affected by the static and dynamic cutting forces, the particular combination of work material/tool material, the variations in their batch qualities and the tool wear.

Errors in form can result in tapers, lack of straightness, convex or concave shapes, deviation in parallelism, errors in angularity, waviness, etc. Usually a mixture of various form errors is present, as shown in the example of a cylindrical component in Fig. 6.2. The upper and lower limits of size are superimposed as would be required in order to determine the cumulative errors which are present.

6.1   Component measuring techniques

One of the difficulties in evaluating the accuracy of machine tools based upon performance tests lies in the actual measurement of the testpieces. Numerically controlled machine tools produce such a low work inaccuracy today that the errors are of the same order as those permissible in conventional measuring tools, e.g. external micrometers, micrometer gauges, caliper gap gauges, etc.

For workpieces which are of a relatively simple form, this problem is overcome by employing multi-measurement inspection fixtures. In the area of gear-cutting manufacture, the precision measurement of worms and bevel

Measuring machine

Technical data

Measurement X = 500 (mm)
range Y = 200 (mm)
Z = 300 (mm)

Largest work capacity
X = 1100 (mm)
Y = 400 (mm)
Z = 400 (mm)

Maximum permissible mass of work : 150 kg
Resolution : 0.5 μm
Maximum error : $\pm$ (0.8 + L / 250) μm
L = length being measured (mm)

Stylus diameters : 0.5 to 15 mm

Stylus forces : 0.1 N , 0.2 N , 0.4 N

Work temperature range
of electronics 0 to 40°C

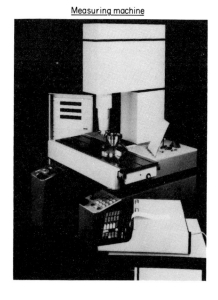

Fig. 6.3   Technical data of a measuring machine (Zeiss UMM 500)

gears was until recently an unsolved problem. A decisive advance towards the solution of this difficulty has been the development of three-dimensional measuring machines which exhibit a very fine resolution and are capable of very high accuracy in measurement. An example of such a machine is shown in the photograph in Fig. 6.3, together with its most important technical data. The stylus which is used to contact the surfaces being measured is able to reach every point within the space where measurements are taken. A computer unit attached to the machine prepares the theoretical data, determines the positioning for a given measuring activity, records the actual measured coordinates and processes the output data.

To provide a hard copy of the measurement results measuring machines are fitted with graphical plotters and high-speed printing terminals. Figure 6.4 shows as an example the measurement of a trochoid-shaped cylinder block for a Wankel motor. The actual profile of the curve is produced on the measuring machine by a series of dots with spacings which may be pre-set; this is then compared with the theoretical profile produced by the computer. The plotted diagram shows the form errors to a magnified scale.

The theoretical shape or reference form may be obtained by passing a stylus around a 'master form' and storing it in the computer or it may be generated by a simulation process. Figure 6.5 shows an example of a model used for calculating the theoretical contour of bevel gear teeth, which is based upon a simulation of the generating process. All setting and motion possibilities of the machine are taken into account. By step-by-step changes in the angles α_{12}, α_{45} and α_{89} the motions of the workpiece, cutter and machine are

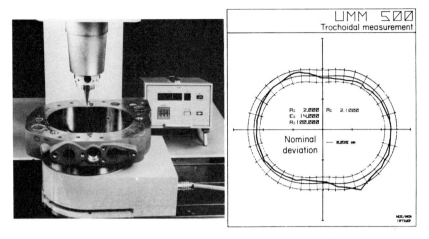

Fig. 6.4   Automatic measurement of a trochoidal profile (Zeiss)

described for any given instant. The result of this procedure is to produce the flank contour of the gear teeth in the form of a series of discrete points as a reference for the measurement process. This is achieved by producing a network of measuring points on the three-dimensional contour of the tooth flank to be measured with the corresponding theoretical values as obtained from the simulation of the machining process shown in Fig. 6.5.

To enable the measured values to be recorded a drawing of the theoretical contour of the gear-tooth flank is produced as shown in Fig. 6.6. The measurement points of the gear-tooth flank are chosen in such a manner that they

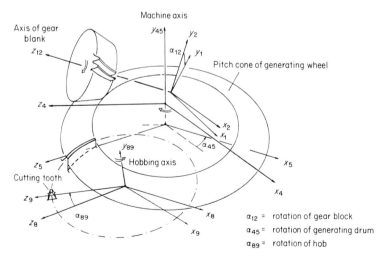

Fig. 6.5   Co-ordinate diagram for simulating the cutting process of bevel gears

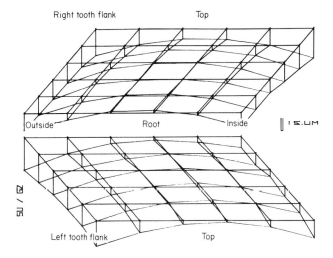

Fig. 6.6   Bevel gear-tooth measurement with a three-
coordinate measuring machine

may be associated to the top/root or outside/inside of the tooth with parallel
lines. The computer records the measured deviations to a magnified scale on
this network, indicating the particular over- or under-size measured values
above or below the corresponding theoretical values. By employing this
method of recording of the measured values it becomes possible to identify
the contour deviations on the gear-tooth flank notwithstanding the comparatively crude network of the positions of the individual measuring points.

6.1.1   Inspection of geometry and positional uncertainty

In this section we describe typical standard testpieces which are used for
evaluating the work accuracy of some machine tools. Specialist committees of
the Association of German Engineers (Verein Deutscher Ingenieure—VDI)
and the German Society for Quality Control (Deutsche Gesellschaft für Qualität—DGQ) concern themselves with the design of standard testpieces and
standard specifications (statistical evaluation techniques). The results of their
work are published in the joint guideline VDI/DGQ 3441 ff.[3] for various
types of machine.

Figure 6.7 shows a testpiece for an acceptance test of lathes (VDI/DGQ
Guideline No. 3442).

Number of prescribed testpieces $n = 25$ pieces
Material: C45
Finishing cut conditions:
        Depth of cut $a = 0.5$ mm
        Feed rate $s = 0.1$ mm rev^{-1}

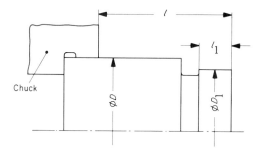

Dimensions (mm)

Swing diameter above bed	≤ 200	>200...315	>315....500
Diameter D of testpiece	40	63	100
Projecting length l of testpiece	40	63	100
Diameter D_1	25....40	>40....63	>63.....100
Minimum length l_1	16	16	16

Note: In special cases (e.g. automatic screw machines)
the design of the testpiece is a matter for agreement
between manufacturer and customer

Fig. 6.7 Testpiece for acceptance tests of lathes
(VDI–DGQ)

After the diameter d_1 has been turned for a length l_1 the deviation of the diameter d_1 from the set dimension on the lathe is measured, having machined it under 'finishing-cut' conditions, during which there is no significant static load. Hence only the geometry and positional accuracy of the machine is being examined.

A testpiece suitable for performance tests of milling machines is shown in Fig. 6.8. It is required to be milled at various positions on the machine table.

Number of testpieces for each table position $n = 5$
Work material C45 or GG25
Finishing depth of cut $a = 0.5$ mm
Feed per cutting edge $s_z = 0.05$ mm rev^{-1}
Cutter diameter to be the maximum suitable for the spindle or arbor and
placed near the front bearing
Number of cutting edges may be according to choice
Approach angle $\kappa = 90°$

The dimensional errors on the testpiece in the x, y and z co-ordinates are determined in relation to the theoretical setting and the position on the table.

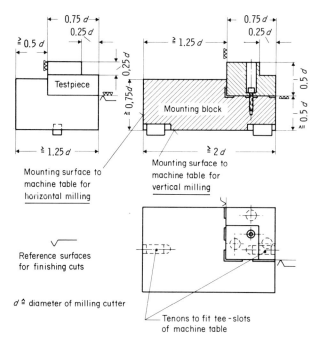

Fig. 6.8 Testpiece for acceptance tests of milling
machines

Figure 6.9 illustrates a testpiece for numerically controlled drilling
machines. The order in which the holes are drilled must be according to the
path indicated on the diagram.

Hole diameter d = 25 to 40 mm
Depth of hole = $2d$

The positional errors and the backlash are determined by the differences
which are measured in dimensions a, b, c and d.
Figure 6.10 shows a picture and drawing of a testpiece for continuous-path
controlled milling machines. The following criteria can be inspected on such a
testpiece for both up-cut (conventional) and down-cut (climb milling):

(a) backlash, from the differences in dimensions a and b, c and d (positional accuracy);
(b) parallelism between machined faces (geometry);
(c) angular errors (control system, interpolation);
(d) errors in circular form (control system, interpolation, play);
(e) dimensional errors (control system, feed-back/measuring system, static
deformations, temperature effects, wear).

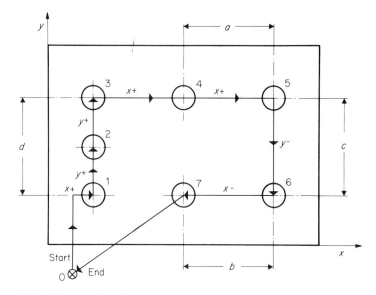

Fig. 6.9   Testpiece for performance tests of NC drilling machines

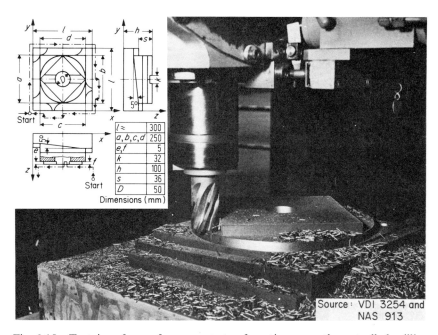

Fig. 6.10   Testpiece for performance tests of continuous-path controlled milling
machines

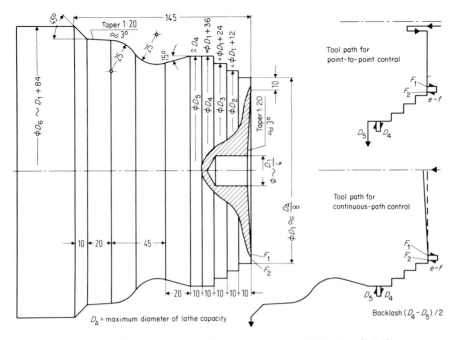

Fig. 6.11   Testpiece for performance tests on NC lathes (VDI)

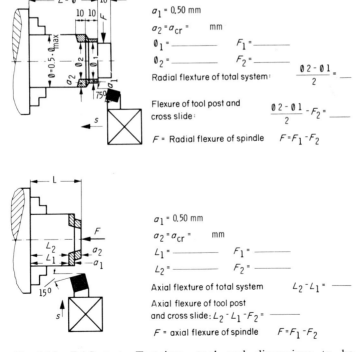

Fig. 6.12   BAS test: Testpiece, tool and dimensions to be inspected for chucking lathes

The testpiece shown in Fig. 6.11 enables the following criteria to be examined on continuous-path or point-to-point controlled lathes:

(a) backlash, from the dimensions e and f, d_4 and d_5 (positional accuracy);
(b) angular errors, path errors (only for continuous-path control) (control system, interpolation);
(c) dimensional errors (control system, static deformation, temperature effects, wear).

6.1.2 Tests for static stiffness

The testpiece shown in Fig. 6.12 (from a Swedish standard recommendation BAS) may be used to determine the effect of static flexibility upon the work accuracy. The testpiece is firstly machined to a cutting depth a_1. Due to the stepped-diameter shape of the testpiece, the machine is then subjected to a greater load corresponding to the cutting depth a_2 (established previously during chatter tests). The differences in diameter and length indicate the total flexure of the machine–workpiece–tool system. By measuring the flexure (F) between the spindle and the lathe bed, the deformation which is caused by the spindle, headstock and bed can be established. The flexure of the tool post and cross slide may also be determined (see Fig. 6.12).

7

LIMITING CUTTING CAPABILITY (DYNAMIC BEHAVIOUR)

The establishment of a limiting cutting capability by measurement presupposes that the point of change from a stable to an unstable machining process can be precisely identified. The criteria which are applied for defining such a limiting cutting capability are the so-called limiting critical chip width b_{cr} or the limiting depth of cut a_{cr} at which it is still just possible to machine under stable conditions.

In order to find these limiting values the depth of cut is gradually increased stage by stage until vibrations of high amplitudes suddenly occur. Generally these are also detected by a noise, the frequency of which is mainly governed by the natural frequency for the introduction of chatter. Particular care must be taken to detect very low frequency vibrations (e.g. 'gate' vibrations) which cannot be sensed by the human ear. The waviness of the surface finish of the

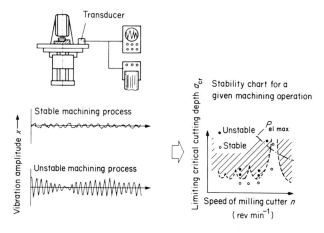

Fig. 7.1 Establishment of a limiting critical cutting depth and presentation of results on a stability chart

v_{30} = Cutting speed which permits a
30 min tool life at a given feed rate
(tool-life criterion VB = 0.4 mm)

Limiting cutting criteria
b_{cr} = limiting critical chip width

		Cutting conditions						Acceptance values			
		s, mm	v, m min^{-1}	d_w	κ_M	L	Tool	P_{max}, kW	P_{lim}, kW	Classi-fication	b_{cr}, mm
A		0.002 and 0.004 d_{cr}	v_{30}	0.6 d_{cr}	45°	0.3 d_{cr}	HM P 30				
B		$\dfrac{0.1}{0.2}$	100	0.4 d_{cr}	90°	0.5 d_{cr}	HM P 30			0 A B C	< 0.06 d_{cr} < 0.04 d_{cr} > 0.02 d_{cr} < 0.02 d_{cr}

Fig. 7.2   UMIST tests: cutting conditions and acceptance values for chucking lathes

workpiece is a second indicator which can be used in all cases. The most positive technique for the direct measurement of vibrations has been found to be by the use of linear, velocity or acceleration transducers which are mounted near the cutting point (e.g. on the spindle or machine table); the signals of these are recorded on an oscillograph or pen-recorder (Fig. 7.1).

Currently there are no standard specifications for establishing limiting cutting capability. However, a number of institutes and institutions give recommendations for test conditions.

Figure 7.2 shows some conditions for acceptance tests for chucking lathes. Two machining processes are chosen as examples which are to be examined under two differing boundary conditions. Figure 7.3 shows recommended test conditions for the inspection of boring and milling machines. Various tool/workpiece configurations are prescribed for different arrangements of machine components. (These test recommendations are taken from a British recommendation.[38])

The test conditions of the Swedish companies Bofors, Alfa-Laval, Asea and Scania Vabis (BAS specification), the ENIMS machining tests developed in the USSR and the VDF machining tests of the company Vereinigte-Drehmaschinen-Fabriken (United lathe manufacturers) all have similar recommendations for the inspection of limiting cutting capability of metal-cutting machine tools.[38,39]

Some critical comments may be made with respect to these types of machining tests, because in general the recommended testing techniques do not fully satisfy the demands of a machine-tool evaluation. The repeatability of a

I. Face milling with spindle fully retracted

Diameter of spindle d_s	Diameter of milling cutter d_c	Remarks
63 mm	200 mm	Cutter mounted directly on to spindle flange with a standard adapter fitted to face mill
70, 75 mm	250 mm	
90, 10 mm	315 mm	

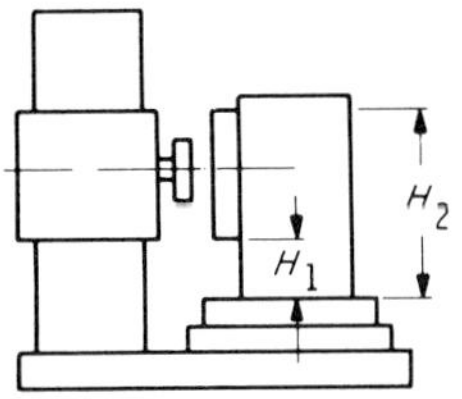

Test	H_2	H_1	v, m min^{-1}	s, mm	Remarks
A_1	$0.6\,H_{max}$	$0.3\,H_{max}$	150	0.25	P_{max} for the direction of feed a,b,c,d,e,f
A_2	$H_{max}-d_F$	$0.6\,H_{max}$	150	0.25	$P_{gr} \geqslant 0.75\,P_{max}$ for the direction of feed a,b,c,d,e,f

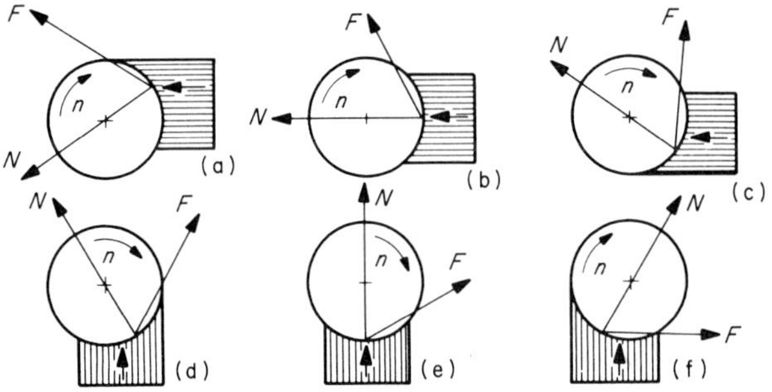

Fig. 7.3 UMIST test: test conditions for horizontal boring and horizontal milling machines I

machining test is very limited due to the unavoidable influence of the cutting process itself. Apart from tool wear, the variations in the work material constituents and the inevitable changes in the technological conditions have an almost uncontrollable influence upon the test results.[13,22] Moreover, the value of the test results must be questioned as they are only applicable to the particular case which is examined at a given time. Inspections which consider all the relevant machining possibilities with respect to varying tool/workpiece configurations are completely impractical due to the high costs which would be involved.

The effort which is necessary to determine the minimum value for the limiting cutting depth for just one single tool/workpiece configuration when milling may be clearly seen from Fig. 7.4. Due to the relationship between the limiting critical chip width and the rotational speed of the milling cutter it is advisable at the very least to establish the chatter curve accurately.

The application of machining tests for an evaluation of the dynamic behaviour of machine tools is, as a result of the limitations described above,

138

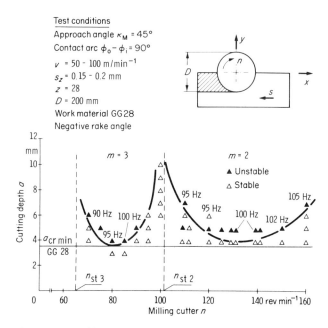

Fig. 7.4   Stability chart of a milling machine for one
tool/workpiece configuration

only recommended for special-purpose machines or for machines on which the number of variations in their application is limited. In contrast, the technique described in section 4.9 which is based upon dynamic flexibility measurements and subsequent machining simulations is very effective because of its wide-ranging output of results of reasonable inspection costs and easy repeatability of the readings obtained.

8

SUMMARY

The purpose of this volume *Metrological Analysis and Performance Tests* has been to provide a comprehensive appreciation of the various methods available for the determination and evaluation of the characteristics of machine tools. Apart from the geometric and kinematic features of machines, their thermal, static, dynamic and acoustic aspects have also been dealt with.

The aim has been to introduce techniques which will permit errors in machine tools to be determined with a view to aid their correction on the one hand and to give an appraisal of a machine being delivered to the customer on the other.

The current development of the measurement technology has been presented in detail and, where available, appropriate guidelines and standard specifications have been dealt with within which testing and acceptance techniques are described.

In addition, methods are discussed in connection with the metrological techniques which enable the weak points of the machine to be identified with respect to the individual characteristics, so that positive steps may be taken towards an improvement.

Parallel with this identification and evaluation of the machine characteristics by direct measurement, the last chapter covers the indirect approach to the quality of the machine, i.e. by the application of performance or acceptance tests and the subsequent inspection of the finished test pieces. In this connection the value of such tests with respect to the possibility of an analysis of the individual features of the machine is critically examined.

REFERENCES

1. DIN 8601 (Draft): *Acceptance Conditions for Machine Tools (General Rules)*, Deutscher Normenausschuss (April 1975).
2. DIN 8605–8668 (Draft): *Abnahmebedingungen für die einzelnen Maschinenarten* (DIN 8605–8613); *Drehmaschinen* (DIN 8615–8623); *Fräsmaschinen* (DIN 8625–8626); *Bohrmaschinen* (DIN 8630–8635); *Schleif und Honmaschinen* (DIN 8642); *Verzahnmaschinen* (DIN 8660–8668); *Hobel-, Stoss- und Räummaschinen*), Deutscher Normenausschuss (April 1975).
3. VDI/DGQ 3441–3445: *Statistical Examination of Exact Working and Positioning of Machine Tools*, Beuth–Verlag, Cologne and Berlin (1977).
4. DIN 45635: *Measurement of Airborne Noise Emitted by Machines*, Deutscher Normenausschuss (1972).
5. J. Tlusty: Richtlinien und Vorschriften für die Prüfung von Werkzeugmaschinen, *CIRP–Annalen*, **XI**, No. 3, 125 (1963).
6. Th. Stöferle, F. Ertl, P. A. McKown, A. J. Scarr and M. Weck: Specifying the accuracy of multi-axis measuring machines and machine tools, CIRP–Work-Shop, CIRP–Secretariat General, 19 Rue Blanche, Paris 1976–77.
7. J. Tlusty; Testing and evaluating the accuracy of numerically controlled machine tools, *Transactions of the Society of Manufacturing Engineers*, Dearborn, Michigan, USA (1973).
8. Tilo Pfeiffer: Neuere Messverfahren zur Beurteilung der Arbeitsgenauigkeit von Werkzeugmaschinen, Habilitation, Technische Hoschschule Aachen (1972).
9. Tilo Pfeiffer: Untersuchung der geometrischen und kinematischen Genauigkeit von Werkzeugmaschinen unter Einsatz neuer Laser-Technologien, DFG–Forschungsbericht Op/1/157 (1975).
10. J. Peters and P. Vanherck: An axis of rotation analyser, in *Proceedings of the Fourteenth International Machine Tool Design and Research (MTDR) Conference, Manchester, 1973*, Pergamon Press, Oxford (1973).
11. *Schwingungen an spanenden Werkzeugmaschinen*, Handbuch zur Aachener VDW–Konstrukteur–Arbeitstagung, Technische Hochschule, Aachen (1971).
12. O. Danek, M. Polacek, W. Spacek and J. Tlusty: *Selbsterregte Schwingungen im Werkzeugmaschinenbau*, VEB–Verlag Technik, Technik, Berlin (1952).
13. Manfred Weck and Klaus Teipel: *Dynamisches Verhalten spanender Werkzeugmaschinen. Einflussgrössen, Beurteilungsverfahren, Messtechnik*, Springer–Verlag, Berlin, Heidelberg, New York (1977).
14. J. Milberg: Analytische und experimentelle Untersuchungen zur Stabilitätsgrenze bei der Drehbearbeitung, Dissertation, Technische Universität Berlin (1971).
15. E. Rehling: Entwicklung und Anwendung elektro-hydraulischer Wechselkrafterreger zur Untersuchung von Werkzeugmaschinen, Dissertation, Technische Hochschule Aachen (1965).

16. K. Beckenbauer: Entwicklung und Einsatz eines Aktiven Dämpfers zur Verbesserung des dynamischen Verhaltens von Werkzeugmaschinen, Dissertation, Technische Hochschule Aachen (1970).

17. Tilo Pfeifer: Berührungsloser elektromagnetischer Schwingungserreger für dynamische Untersuchungen an Werkzeugmaschinen, Dissertation Technische Hochschule Aachen (1968).

18. Manfred Weck and Ernst-Kurt Prössler: Untersuchung des dynamischen Verhaltens spanender Werkzeugmaschinen und deren einzelner Bauelemente und Kopplungsstellen mit Hilfe aperiodischer Testsignale, DFG–Forschungsbericht We 550/9 (June 1977).

19. Hottinger Baldwin Messtechnik, Darmstadt: Die DMS–Technik, Daten und Hilfsmittel, Firmenzeitschrift (1969).

20. Hewlett and Packard, Santa Clara (USA): *Software Manual for Fast–Fourier–Analyser–Systems*, Service manual for HP Fast–Fourier–Analysator 4551 B.

21. Ernst-Kurt Prössler: Modalanalyse—ein neuartiges und effektives Hilfsmittel zur Darstellung des dynamischen Maschinenverhaltens, *Industrie-Anzeiger*, **98**, No. 42 (1976).

22. Klaus Teipel: Beurteilung der dynamischen Nachgiebigkeit spanender Werkzeugmaschinen—Entwicklung und Anwendung eines Verfahrens auf der Basis gemessener Nachgiebigkeitsfrequenzgänge, Dissertation, Technische Hochschule Aachen (1977).

23. W. Kraag and H. Weissing: *Schallpegelmesstechnik*, VEB–Verlag Technik, Berlin (1970).

24. G. Kurtze: *Physik und Technik der Lärmbekämpfung*, Verlag G. Braun, Karlsruhe (1964).

25. I. I. Slawin: *Industrielärm und seine Bekämpfung*, VEB–Verlag Technik, Berlin (1960).

26. M. Heckel and H. A. Müller: *Taschenbuch der technischen Akustik*, Springer Verlag, Berlin, Heidelberg, New York (1975).

27. DIN 45633: *Sound Level Meters, General Requirements*, Part 1, Deutscher Normenausschuss (March 1970).

28. Unfallverhütungsvorschrift (UVV) 'Lärm', VBG 121, Karl Heymanns–Verlag, Cologne (1974).

29. DIN 45641: *Averaging of Time-varying Sound Levels, Rating Levels*, Deutscher Normenausschuss, (June 1976).

30. VDI 2058: *Assessment of Working Noise in the Vicinity*, Part 1, Verein Deutscher Ingenieure (June 1973).

31. DIN 45667: *Classification Method for Evaluation of Random Vibrations*, Deutscher Normenausschuss (October 1969).

32. DIN 45651: *Octave Filters for Electron Acoustical Measurements*, Deutscher Normenausschuss (January 1964).

33. DIN 45652: *Third Octave Filters for Electron Acoustical Measurements*, Deutscher Normenausschuss (January 1964).

34. S. Lachenmaier: Überlegungen zur Geräuschanregung in Zahnradgetrieben. Bericht über die 19. Arbeitstagung Zahnrad- und Getriebeuntersuchungen, 14–19 Oct. 1976, WZL der TH Aachen.

35. Manfred Weck: Geräuschemission von Werkzeugmaschinen–Forderungen und wirtschaftliche Grenzen, ICM Tagungsbroschüre, Hannover 1977, Verein Deutscher Werkzeugmaschinenfabriken e.V., Frankfurt.

36. Wilfried Melder: Geräuschemission spanender Werkzeugmaschinen—Einflussgrössen, Beurteilungsverfahren, Messtechnik, Dissertation, Technische Hochschule Aachen (1976).

37. Manfred Weck: Geräuschemission von Drehmaschinen. BMA/VDW–

Forschungsbericht, TH Aachen, July 1977, Wirtschaftsverlag Nordwest GmbH, Bremerhaven (July 1977).

38. J. Tlusty and F. Koenigsberger: University of Manchester, Institute of Science and Technology (UMIST): Specifications and test of metal-cutting machine tools, in *Proceedings of the nineteenth and twentieth Conferences*, February 1970, Vols. 1 and 2, Revell & George Ltd., Manchester (1970).

39. AB Bofors, Alfa-Laval AB, Asea u. Saab-Scania: Bearbeitungstests zur Untersuchung des dynamischen Maschinenverhaltens der Firmen AB Bofors, Alfa-Laval AB, Asea und Saab-Scania (BAS-Norm), Sweden (1970).

INDEX

143